Self-Interest
vs.
Altruism
in the Global Era

How society can turn
self-interests into mutual benefit

Self-Interest vs. Altruism in the Global Era

How society can turn
self-interests into mutual benefit

LAITMAN
KABBALAH PUBLISHERS

Michael Laitman, PhD

SELF-INTEREST VS. ALTRUISM IN THE GLOBAL ERA
How society can turn self-interests into mutual benefit

Published by Laitman Kabbalah Publishers
www.kabbalah.info   info@kabbalah.info
1057 Steeles Avenue West, Suite 532, Toronto, ON, M2R 3X1, Canada
2009 85th Street #51, Brooklyn, New York, 11214, USA

Printed in Canada

ISBN 978-1-897448-65-6

Library of Congress Control Number: 2011910435

Professional Consultation: Prof. Valeria Khachaturian,
Prof. Itzhak Orion, Dr. Yael Sanilevich, Dr. Eli Vinokur, and
Mr. Ronen Avigdor
Copy Editor: Claire Gerus
Associate Editors: Alicia Goldman, Brad Hall,
Charles Bowman, Dan Berkovitch, Eric Belfer,
Gilbert Marquez, James Torrance, Keren Applebaum,
Noga Burnot, Rachel Branson, Riggan Shilstone, and
Tom Dorben.
Layout: Baruch Khovov
Cover: Rami Yaniv
Producer: Leah Goldberg
Executive Editor: Chaim Ratz
Producer and Publisher: Uri Laitman

FIRST EDITION: MARCH 2012
FIRST PRINTING

CONTENTS

About the Author

Professor Michael Laitman was born in 1946 in Vitebsk, Belarus. He received his Professorship in Ontology and the Theory of Knowledge, and his PhD in Philosophy and Kabbalah from the Moscow Institute of Philosophy at the Russian Academy of Sciences. He received his MSc in Medical Cybernetics from St. Petersburg State Polytechnic University.

Prof. Laitman began his career conducting research at the St. Petersburg Blood Research Institute, specializing in electromagnetic regulation of the blood supply in the heart and brain.

His scientific career took a sharp turn in 1974 when he immigrated to Israel. There, he worked for the Israeli Air Force for several years before becoming self-employed.

In 1976, he began his studies of Kabbalah, which he continues to research to this day. In 1979, he came across Kabbalist Rav Baruch Shalom HaLevi Ashlag (1906-1991), known as "Rabash." Rabash was the first-born son and successor of Kabbalist Rav

Yehuda Leib HaLevi Ashlag (1884-1954), known as "Baal HaSulam" for his *Sulam* (Ladder) commentary on *The Book of Zohar*. Rabash followed his father's scientific approach in the teaching of Kabbalah, which suited Laitman very well.

After Rabash's demise in 1991, Laitman began to teach what he had learned from his teacher, passing on the methodology of Baal HaSulam.

Laitman firmly believes that Kabbalah holds the solution to the current world crises. For this reason, he asserts that Kabbalah belongs to all of humanity, and not to any single person or group of persons. Therefore, he insists on teaching Kabbalah to everyone, regardless of age, sex, nationality, or race.

To date, Prof. Laitman has been teaching Kabbalah for over thirty years, relying on the writings of Baal HaSulam and Rabash. He has already published more than thirty books on Kabbalah, among which are *The Zohar: Annotations to the Ashlag Commentary*, *Kabbalah, Science and the Meaning of Life*, *From Chaos to Harmony*, *Attaining the Worlds Beyond*, and *Kabbalah Revealed*.

His website, *www.kabbalah.info*, is the largest source of authentic Kabbalah on the web, offering authentic Kabbalah materials in over 30 languages, and free introductory courses in Kabbalah. His daily lessons are broadcast live and free of charge on Israeli TV, as well as on *www.kab.tv* in English, Spanish, Russian, German, French, Turkish, German, and Hebrew.

Editor's Note

The writing of *Self-Interest vs. Altruism in the Global Era*, which aims to explain the evolution of existence originally seemed an unrealistic objective. It appeared to be especially challenging in light of the scientific, as well as Kabbalistic, angle the book was to take. And yet, with the help of many good friends from around the world, we have successfully completed the task.

I feel it is my duty to thank each of those who contributed to the effort, and I apologize if anyone has been inadvertently omitted. In my view, this book is the creation of a group, rather than the work of a few proficient individuals, and this is where its merit lies.

Below is the list (in order of the alphabet) of people who contributed their time, effort, and often money to the completion of *Self-Interest vs. Altruism in the Global Era*:

Research and proofreading (usually both): Anastasia Cherniavski, Annabelle Fogerty, Asaf Ohayon, Asta Rafaeli, Avraham Cohen, Beth Shillington, Christiane Reinstrom,

Crystlle Medansky, Daniel Lange, Eli Gabay, Geoffrey Best, James Torrance, Jonathan Libesman, Julie Schroeder, Kimberlene Ludwig, Loredana Losito, Markos Zografos, Marlene Bricker, Michael R. Kellogg, Michal Karpolov, Pete Matassa, Peter LaTona, Sandra Armstrong, Shari L. Kellogg, Veronica Mengana, Yehudith Sabal, and Zhanna Allen.

Editors: Alicia Goldman, Brad Hall, Charles Bowman, Dan Berkovitch, Eric Belfer, Gilbert Marquez, James Torrance, Keren Applebaum, Noga Burnot, Rachel Branson, Riggan Shilstone, and Tom Dorben.

Production: Leah Goldberg

Administration: Avihu Sofer, Alex Rain (images).

I'd also like to extend my deepest gratitude to Prof. Valeria Khachaturian, Prof. Itzhak Orion, Dr. Yael Sanilevich, Dr. Eli Vinokur, and Mr. Ronen Avigdor for reviewing the scientific, historic, and financial information presented in the book.

And last but certainly not least, kudos to Baruch Khovov, the designer who always delivers, to Rami Yaniv, who responded to a last-minute request to design the cover, to Claire Gerus, an editor whose work has been an emblem of quality to me, and of course, to Uri Laitman, our publisher, whose dedication and diligence are an inspiration to all who work with him.

Sincerely,
Chaim

FOREWORD

I suppose all children go through a period of asking the "big" questions. Mine were, "Where do we come from?" "Where do we go when we are no longer here?" and especially, "What is the purpose of life?" Perhaps it is because both my parents were doctors that I felt naturally inclined to seek the answers in science. And perhaps because I was searching in science, the answers I found were of a more inclusive and general nature.

My science of choice was cybernetics—medical bio-cybernetics, to be exact. This was to be my research tool. At the time, cybernetics was a new and innovative field of research, enabling scientists to explore complex systems and to find the mechanisms controlling them. I was particularly interested in the human body and its control systems. Through cybernetics, I had endeavored to unlock the secret of human existence itself: the body and the soul that (so I believed) inhabited it.

But my hopes were thwarted. Yes, science had taught me a great deal about life, or rather, about how a new life begins and how it is sustained. Yet, it taught me nothing about the more fundamental questions of meaning that drove my research: what is life and what is it for?

The craving to decipher life's meaning kept me on my toes, probing every shred of data I could find. I continued my search in science, philosophy, and even religion until I gained a plethora of new knowledge and understanding of life. But just as with my initial experience with cybernetics, none of these seemed to address my deepest questions of meaning and purpose.

And then one day, I suddenly reached the conclusion of my lengthy quest when I unexpectedly came across what I later discovered was a science called, "Kabbalah." In retrospect, no part of my search had been redundant or regrettable. Science, philosophy, and religion were all necessary "stops" on my way to Kabbalah, though I never really stopped at any of them. Each of them contributed to my understanding of life's meaning and the purpose of human existence, and each now takes its rightful place in the whole, and (might I add) wholesome worldview that Kabbalah helped me establish.

Moreover, I discovered a connection between the purpose of human existence and the multiple global crises that the world now faces. Through Kabbalah, I acknowledged the inevitability of these crises, their inevitable resolution in peace and prosperity, and the free choice that we have in *how* we resolve them—by collaborating and cooperating, but mostly by becoming aware of our unity and interdependence. More than anything, I discovered that the ancient Kabbalistic concepts on human relations provide a platform on which to build viable societies that promote such amicable relations.

The concept that the current global threats are preordained is not my own. Neither is the idea that the crises are a springboard to a reality that exceeds our wildest dreams. Both notions have existed for millennia, but have only now begun to surface because it is the first time that a necessary, twofold condition has been met: people are desperate enough to seek a solution, and a clear enough explanation of that solution is available. As for my role in the unfolding of these concepts, it is to serve as presenter and facilitator. Yet, as much as I believe in the validity of these ideas, by no means do I claim proprietary rights to them. They are solutions and ideas that I have learned from my teachers throughout the years.

As I hope to show in the chapters ahead, contemporary science and modern thinking now make it possible to meet these conditions and to unveil the age-old paradigm explained in the science of Kabbalah. Thanks to quantum physics, which dared to challenge the Newtonian paradigm of reality, we can deem such concepts as "the oneness of reality" worthy of consideration. And thanks to philosophy, which devoutly cultivated the idea of free thought, we can now share ideas and learn from one another.

Hence, while the concepts I am about to introduce are entirely Kabbalistic, I will show that many of them parallel with modern science. It is my hope that in the spirit of pluralism, they will be met with an open mind and an open heart. And if the views presented here invoke contemplation in even one reader, I will be fully rewarded.

Michael Laitman

INTRODUCTION

At the time these words are written, the world is still reeling from the longest recession since the Second World War. Tens of millions of people throughout the world have lost their jobs, their savings, their homes, but most important—their hopes for the future.

Our health, it seems, is not more wholesome than our wealth. Modern medicine, the pride and joy of Western civilization, is grappling with resurfacing diseases previously believed to be extinct. According to a report published by the Global Health Council, "Diseases once believed to be under control have re-emerged as major global threats. The emergence of drug-resistant strains of bacteria, viruses and other parasites poses new challenges in controlling infectious diseases. Co-infection with multiple diseases creates obstacles to preventing and treating infections."[1]

Earth, too, is not as hospitable as before. Books such as James Lovelock's *The Revenge of Gaia*,[2] Ervin Laszlo's *The Chaos Point*,[3] and films such as Al Gore's *An Inconvenient Truth* are just three examples of a cavalcade of alarming reports on Earth's deteriorating climate.

As global warming melts the ice caps in the poles, sea levels rise. This has already caused dramatic shifts and tragic events. A report by Stephan Faris in *Scientific American*[4] lists some of the places already affected by climate change. In Darfur, clashes between nomadic and sedentary tribes that broke out due to a decades-long drought escalated into a rebellion against the Sudanese government's neglect. Subsequently, the crisis spilled over into Chad and the Central African Republic.

Also in that report, the Pacific island nation of Kiribati declared its lands uninhabitable and asked for help in evacuating its population. In March, 2009, Peter Popham, a writer for *The Independent*, provided another angle to the climate predicament: "Global warming is dissolving the Alpine glaciers so rapidly that Italy and Switzerland have decided they must re-draw their national borders to take account of the new realities."[5]

A more tragic result of climate change is hunger, caused by extended droughts in some areas and constant flooding in others. According to the World Food Programme, nearly a billion (1,000,000,000) people worldwide are constantly hungry. Worse yet, in excess of nine million (9,000,000) people die every year from hunger and related causes, more than half of whom are children.[6] This means that today, in the most technologically advanced era in the history of humankind, a child dies every six seconds due to lack of food and water.

In our homes, problems abound, as well. *The New York Times*[7] announced that according to a census released by the American Community Survey, divorce rates have risen to the point that today there are more unmarried couples in America than married ones. It is the first time in history that single-parent families are the norm, and double-parent ones are the exception.

Many scientists, politicians, NGOs, and UN related organizations warn that humanity is facing a risk of unprecedented catastrophes on a global scale. Anything from mutated avian flu through nuclear war, to a massive earthquake could wipe out millions and drive billions into destitution.

Yet, crises have been occurring throughout history. Our era is not the first in which humanity has been at risk. The Black Death pandemic of the 14th century and the two World Wars easily outweigh the peril that our current plight presents. Nevertheless, what distinguishes the current crisis from those previous is the tension characterizing the current state of humanity. Our society has gone to the extreme in two directions that seem to conflict with one another—globalization and the interdependence it entails on the one hand, and increasing alienation and personal, social, and political narcissism on the other. And that is a recipe for a disaster such as the world has never seen, whether in the financial sector or beyond.

Today, globalization concerns far more than financial interdependence. We have become globally interconnected in every realm of life: the computers and TVs we use to entertain ourselves come (primarily but not exclusively) from China, Taiwan, and Korea. The cars we drive are assembled (again, primarily) in Japan, Europe, and the U.S., but their parts are made in numerous other countries. The clothes we wear often arrive from India and China, while the food in our refrigerators comes from all over the world.

What's more, throughout the world people watch Hollywood films and learn English by the millions. In fact, of the approximately 1.4 billion English speakers worldwide, only 450 million are native English speakers, and China alone produces over twenty million new English speakers each year, reports the *Asia Times* in a September 15, 2006 story titled, "'Native English' is losing its power."[8]

On March 8, 2009, Wachovia Corp. economist, Mark Vitner, gave a rather palpable description of the world's globalized situation when he described the interconnectedness of the credit markets on MSNBC: "It's like trying to unscramble scrambled eggs. It just can't be done that easily. I don't know if it can be done at all."[9]

But the problem with globalization is not only that it makes us interconnected; it also makes us *interdependent*, and instead of using these interconnections to thrive, we become engaged in a constant tug of war. What would happen to the oil-rich countries if the world suddenly shifted to wind and solar energy? What would happen to America if China stopped buying dollars? What would happen to China, Japan, India, and Korea if Americans had no dollars with which to buy Asian-produced goods? And if Western tourists ceased to travel, what would become of the hundreds of millions of people all over the world who provide for their families, thanks to Westerners' hedonism?

Journalist Fareed Zakaria eloquently described this entanglement in a *Newsweek* article titled, "Get Out the Wallets: The world needs Americans to spend": "If I were told by the economic gods that I could have the answer to one question about the fate of the global economy... I would ask, 'When will the American consumer start spending again?'"[10] Indeed, we have become a global village, completely reliant on one another for our sustenance.

Yet, interdependence is only a part of today's complicated picture. While we have been growing increasingly global, we have also become increasingly self-centered, or as psychologists Jean M. Twenge and Keith Campbell describe it, "increasingly narcissistic."[11] In their insightful book, *The Narcissism Epidemic: Living in the Age of Entitlement*, Twenge and Campbell talk about what they refer to as "The relentless rise of narcissism in our culture,"[12] and the problems it causes. They explain that

"The United States is currently suffering from an epidemic of narcissism. ...narcissistic personality traits rose just as fast as obesity." Worse yet, they continue, "The rise in narcissism is accelerating, with scores rising faster in the 2000s than in previous decades. By 2006, 1 out of 4 college students agreed with the majority of the items on a standard measure of narcissistic traits. Today, as singer Little Jackie put it, many people feel that 'Yes siree, the whole world should revolve around me.'"[13] In Webster's Dictionary, narcissism is defined as "egoism," and this, blatantly speaking, means that we have become unbearably selfish.

Thus, our problem is twofold: On the one hand, we are interdependent; on the other hand, we are becoming increasingly narcissistic and alienated. We are trying to lead two ways of life that simply do not meet: interdependence and alienation. Perhaps this is why we spend countless hours chatting with "virtual friends" in online social networks, but are often cold and heartless toward our kin at home. If we were simply interdependent, we would unite, support each other, and be happy. Alternatively, if we were simply selfish, we would part and live by ourselves. But if we are both interdependent *and* selfish, neither way works!

And this, in essence, is the root of the crisis: our interdependence requires us to work together, but our selfishness causes us to deceive and to exploit one another. As a result, the systems of cooperation that we work so hard to build break down, leading to continuing crises.

Hence, the goal of this book is twofold: 1) to shed light on the cause of our interdependence, on one hand, and our self-centeredness, on the other hand; and 2) to briefly outline a feasible *modus operandi* for combining these seemingly conflicting traits to our advantage. To address the first goal, I will explain what I have learned in Kabbalah about the structure of Nature, and particularly, of human nature. To address the second goal, I will combine the ideas of the great 20th century Kabbalist, Yehuda Ashlag, as well as other great Kabbalists, with suggestions from contemporary scientists and scholars from other disciplines.

In the wisdom of Kabbalah, I discovered what I believe to be a viable solution to the current global problems, and I feel grateful that I have been given a chance to present it. It is my hope and, may I say, conviction that through the concepts that Kabbalah offers we can save ourselves, as well as the Big Blue Marble that we live in.

1

MAN'S SEARCH FOR ONENESS

When the worst financial crisis since the Great Depression first broke out in August of 2008, many politicians and financiers in key positions emphasized the need for unity and cooperation. They voiced a need to restrain the egocentric frame of mind dominating Wall Street and expressed a fear of separatist and protectionist tendencies. Headlines such as *The Economic Times'* "World Leaders Seek Unity to Fight Financial Crisis"[14] prevailed in newspapers all over the world, signaling a general willingness to unite and cooperate in the face of economic uncertainty.

At first glance, this spirit is understandable, if not called for. After all, the world's financiers knew that their institutions were linked together so tightly that if one failed, the others would follow, and politicians were warned that if they did not bail out the banks in their countries, their own economies would collapse, precipitating a domino effect that would bring down the entire global economy.

However, in the face of a crisis, it is natural to do the opposite of uniting: close yourself off and protect what is yours. This seems like a safer route than joining forces with "foreigners," especially when those foreigners may be regarded as culprits or, at least, contributors to the making of your plight.

Thus, America—the country generally regarded as the primary perpetrator of the outbreak and rapid escalation of the financial crisis—is not suffering from isolation, because the interconnectedness of the global economy compels economies such as China's to buy dollars and thus provide sustenance to the American economy.

For politicians, it would seem more natural to put their own countries first, as with the British Corn Laws tariffs of the 19th century and President Hoover's 1933 "Buy American" Act. Yet, as the delicate balance of cooperation and self-interest teeters back and forth, we survey the destruction wrought by the financial crisis and find that the majority of voices champion unity and denounce protectionism anIf we consider this question from a purely economic or psychological aspect, we will not arrive at a conclusive answer. However, when we view it from the perspective of the science of Kabbalah, we will see that the forces involved in international relations, and indeed in any relations, are forces of integration, not of isolation. They are far more powerful than any rational or irrational decision-making process, and determine our moves "behind the scenes."

On the international level, these forces determine global trade, politics, treaties, conflicts, and ecology. On the national level, they determine the trends in education, welfare policy, media, and local economy. On the personal level, they determine our relationships with our families, and on the deepest level of existence, they determine evolution—ours and that of every other element in Nature.

When we understand these forces, we will understand why Napoleon, for example, bit off more than he could chew when he tried to conquer Russia, why Hitler did the same (and in the same country, no less), and why Bernard Madoff could not stop until he was stopped. The "bridge-too-far" syndrome is a typical human pitfall, one that the world's greatest leaders and would-be leaders could not resist. Indeed, the forces that make us behave as we do are so much a part of us and of our world that not recognizing them is a risk we mustn't allow ourselves to take.

To understand the forces and elements that create reality and stir it in its course, we must first come to know their origins and their final destinations. Otherwise, trying to understand reality is like trying to grasp the inner workings of a car—its engine, the engine's connection to the gear, the way the gear shifts the power to the wheels, and so on—without explaining that a car is a machine built to transport people safely, comfortably, and quickly from place A to place B. Without explaining the car's purpose, what good is any discussion of its structure?

Like science, Kabbalah researches the inner workings of reality. But unlike science, which observes phenomena and offers theories as to their end goal, Kabbalah sees the goal first and from there explains the structure. That goal, as explained by Kabbalah, is for every person in the world to discover the single, fundamental force that creates and governs all of life. In other words, the goal of Kabbalah is for every person to discover life's creative force, obtain it, and reap all the benefits this discovery implies.

The 20th century Kabbalist, Yehuda Ashlag, known as Baal HaSulam (Owner of the Ladder) for his *Sulam* (Ladder) commentary on *The Book of Zohar*, described Kabbalah and life's purpose in the following way: "This wisdom is no more and no less than a sequence of roots, which hang down by way of cause and consequence, by fixed, determined rules, interweaving to a single, exalted goal described as 'the revelation of the Creator to the creatures in this world.'"[15] Our lives are the vehicle by which to achieve this purpose. Hence, Kabbalists regard our world's physical, historical, and social phenomena as stages toward an end goal, and it is from this perspective that this book will discuss humankind's history and current state.

THE HIDDEN PREVAILING UNITY

Kabbalah is certainly not the only science ever to research Nature's hidden forces that operate our world behind the scenes. According to *The Encyclopedia Britannica*, "Newton's theory of mechanics, known as classical mechanics, accurately represented the effects of forces under all conditions known in his time. ... the theory has since been modified and expanded by the theories of quantum mechanics and relativity."[16] In other words, to make a gross generalization, in the 20th century science was no longer satisfied with Newton's theory because it was insufficient to explain all of Nature's observed phenomena.

In the second half of the 20th century, scientists realized that the new theories, too, fell short of explaining all of Nature's phenomena. This prompted a search for a Grand Unified Theory (GUT). "The dream of theorists [in physics]," according to *The Encyclopedia Britannica*, "Is to find a totally unified theory—a theory of everything, or TOE."[17]

In what seems like a parallel to the quest for the TOE, many prominent theoretical physicists have begun to posit that at the most fundamental level, we—and all parts of reality—are actually one. Pioneering theoretical physicist Werner Heisenberg said, "There is a fundamental error in separating the parts from the whole, the mistake of atomizing what should not be atomized. Unity and complementarity constitute reality."[18]

Heisenberg's contemporary and fellow founder of quantum physics, Erwin Schrödinger, stated in his essay "The Mystic Vision," "The plurality that we perceive is only an appearance; it is not real."[19] Even the great Albert Einstein, in a letter dated 1950, declared, "A human being is part of the whole called by us universe. ...We experience ourselves, our thoughts, and feelings as something separate from the rest, a kind of optical delusion of consciousness."[20]

Yet, proving that all parts of reality are manifestations of a single whole, or developing a TOE that applies to all parts of reality, would require a paradigm that works on all levels of life—physical, mental, and intellectual. And here, physicists are out of their purview. Even the most cutting edge theoretical physicists cannot explain *all* of Nature's observed phenomena.

In particular, a complete explanation of the phenomenon called "consciousness" eludes scientists of all fields. However, consciousness is not only present, but *invariably* affects the results of scientific experiments. In that regard, Dr. Johnston Laurance, former director at the National Institute of Child Health and Human Development, published the following statement in an online essay titled, "Objective Science: an Inherent Oxymoron": "All scientific observation—even at the most fundamental level—is affected by the observer's consciousness. In this regard, the statement 'I'll see it when I believe it,' is more apropos than its commonly stated converse. Numerous studies have shown that consciousness exerts a significant influence on many different endpoints, ranging from bacterial growth to the outcomes of heart patients."[21]

In that essay, Dr. Laurance quoted several other scientists and thinkers who share that view, such as 19th century neurologist Jean Martin Charcot, considered the founder of modern neurology: "In the last analysis, we see only what we are ready to see, what we have been taught to see. We eliminate and ignore everything that is not part of our prejudices."

Thus, if scientific observation affects, distorts, or altogether eliminates the phenomenon being observed, how can science ever be considered 100 percent accurate? Moreover, can any phenomenon be fully understood if at least one key factor of influence—consciousness—is not subject to study and observation?

This is where philosophy steps in to complement science and fill in the gaps of uncertainty. Many great thinkers have done this by expressing the concept of "the oneness of reality." Zeno of Citium, the great 4th century BC Greek philosopher, stated, "All things are parts of one single system, which is called Nature."[22]

Similarly, German philosopher and mathematician, W.G. Leibniz, expressed himself thus in *The Philosophical Writings of Leibnitz*: "Reality cannot be found except in One single source, because of the interconnection of all things with one another."[23]

Certainly, it would be very nice to believe in this perfect picture of oneness, unity, and interconnection among all things. But as eloquent as philosophers may be, a genuine seeker of truth would hardly accept an idea merely because it "sounds" beautiful or true. At the end of the day, the only truly valid test for a theory or concept is one's *personal experience*.

After all, what seems valid and true for one may seem completely false to another. If you project a ray of light through a prism, it will separate the light into all the colors of the rainbow. But if the person you show it to is a monochromat (totally colorblind), it will make no difference what names you give to those shades of gray that he or she will see. To that person, they will all be grays. Similarly, as right as physicists and philosophers may be in their observations on the oneness and indivisibility of reality, to accept this oneness as fact, people must *experience* it for themselves.

While experiencing the oneness of reality may sound mystical to many, the above quotes prove that many proponents of this view are revered scientists, some even Nobel laureates. In fact, the need for a more complete and uniform picture of reality did not arise with the advent of quantum physics, or even with Einstein. Back in 1879, English chemist and physicist, William Crookes, declared, "We have actually touched the borderland where matter and force seem to merge into one another ... I venture to think that the greatest scientific problems of the future will find their solution in this borderland, and even beyond; here, it seems to me, lie ultimate realities, subtle, far-reaching, wonderful."[24]

Indeed, through my search in science in general, and in Kabbalah in particular, I discovered that Crookes' intuition was dead on, because as I explained above, Kabbalah observes the end goal first, and from there explains the structure. And because reality is the vehicle by which to achieve this goal, Kabbalah is inherently a Grand Unified Theory, a Theory of Everything, allowing us to both understand the full scope of reality and to actually experience its oneness.

THE HARBINGER FROM BABYLON

Before we delve into the principles of this Grand (and indeed) Unified Theory called Kabbalah, we should first understand how it originated and give due credit to its "progenitor." Let us, for a moment, journey back through time to ancient Mesopotamia, the cradle of civilization. Roughly four thousand years ago, situated within a vast and fertile stretch of land between the Tigris and Euphrates rivers in what today is Iraq, a city-state called Babel played host to a flourishing civilization. Bustling with life and action, it was the trade center of the entire ancient world.

Babel, the heart of the dynamic civilization we now call "ancient Babylon," was a melting pot and the ideal setting for numerous belief systems and teachings. Its people practiced idol worship of many kinds, and among the most revered people in Babel was a priest named Abraham, who was a local authority in the practice of idol worship, as was his father, Terah.

However, Abraham had a very special quality: he was unusually perceptive, and like all great scientists, he had a zeal for the truth. The great 12th century scholar, Maimonides (also known as the RAMBAM), described Abraham's determination and efforts to discover life's truths in his book, *The Mighty Hand*: "Ever since this firm one was weaned, he began to wonder. ...He began to ponder day and night, and he wondered how it was possible for this wheel to always turn without a driver? Who is turning it, for it cannot turn itself? And he had neither a teacher nor a tutor. Instead, he was wedged in Ur of the Chaldeans among illiterate idol worshippers, with his mother and father and all the people worshipping stars, and he—worshipping with them."[25]

In his quest, Abraham learned what lies beyond the borderland that Crookes described so many centuries later. He found the unity, the oneness of reality that Heisenberg, Schrödinger, Einstein, Leibniz, and others intuitively sensed. In Maimonides' words, "He [Abraham] attained the path of truth and understood the line of justice with his own correct wisdom. And he knew that there is one God there who leads..., and that He has created everything, and that in all that there is, there is no other God but Him."[26]

(To interpret these excerpts correctly, it is important to note that when Kabbalists speak of God, they do not mean it in the religious sense of the word—as an almighty being that you must worship, please, and appease, which in return rewards devout believers with health, wealth, long life, or all of the above. Instead, Kabbalists identify God with Nature, the *whole of Nature*. The most unequivocal statements on the meaning of the term, "God," were made by Baal HaSulam, whose writings explain that God is synonymous with Nature.

For example, in his essay, "The Peace," he writes (in a slightly edited excerpt), "To avoid having to use both tongues from now on—Nature and a Supervisor—between which, as I have shown, there is no difference...it is best for us to...accept the words of the Kabbalists that *HaTeva* (The Nature) is the same...as *Elokim* (God). Then, I will be able to call the laws of God 'Nature's commandments,' and vice-versa, for they are one and the same, and we need not discuss it further.")[27]

"At forty years of age," writes Maimonides, "Abraham came to know his Maker," the single law of Nature, which creates all things. But Abraham did not keep it to himself: "he began to provide answers to the people of Ur of the Chaldeans and to converse with them and to tell them that the path on which they were walking was not the path of truth."[28] Alas, like Galileo after him, and many other great forerunners throughout history, Abraham was confronted by the establishment, which in his case was Nimrod, king of Babel.

Midrash Rabbah, an ancient text written by Hebrew sages in the 5th century C.E., presents a vivid description of Abraham's confrontation with Nimrod, as well as an amusing peek into Abraham's fervor. "Terah [Abraham's father] was an idol worshipper [who also made his living building and selling statues at the family shop]. Once, he went to a certain place and told Abraham to sit in for him. A man walked in and wanted to buy a statue. He [Abraham] asked him, 'How old are you?' And the man replied, 'Fifty or Sixty.' Abraham told him: 'Woe unto he who is sixty and must worship a day-old statue.' The man was ashamed and left.

"Another time, a woman came in with a bowl of semolina. She told him, 'Here, sacrifice before the statues.' Abraham rose, took a hammer, broke all the statues, and placed the hammer in the hands of the biggest one. When his father came, he asked him, 'Who did this to them?' He [Abraham] replied, 'A woman came and brought them a bowl of semolina, and told me to sacrifice before them. I sacrificed, and one said, 'I will eat first,' and the other said, 'I will eat first.' The bigger one rose, took the hammer, and broke them.' His father said, 'Are you fooling me? What do they know?' And Abraham replied, 'Do your ears hear what your mouth is saying?'"[29]

At that point, Terah felt that he could no longer discipline his impertinent son. "He [Terah] took him [Abraham] and handed him over to Nimrod [who was not only king of Babel, but also proficient in the local practices and beliefs]. He [Nimrod] told him, 'Worship the fire.' Abraham responded, 'Should I worship the water, which quenches the fire?' Nimrod replied, 'Worship the water!' He told him: 'Then, should I worship the cloud, which carries the water?' He told him, 'Worship the cloud!'

"He [Abraham] told him: 'In that case, should I worship the wind, which disperses the clouds?' He told him, 'Worship the wind!' He [Abraham] told him, 'And should we worship man, who suffers the wind?' He [Nimrod] told him: 'You speak too much; I worship only the fire. I will throw you in it, and let the God that you worship come and save you from it!

"Haran [Abraham's brother] stood there. He said, 'In any case, if Abraham wins, I will say that I agree with Abraham, and if Nimrod wins, I will say that I agree with Nimrod.' Since Abraham descended to the furnace and was saved, they told him [Haran], 'Whom are you with?' He told them: 'I am with Abraham.' They took him and threw him in the fire, and he died in the presence of his father. Thus it was said, 'And Haran died in the presence of his father Terah.'"[30]

Thus, Abraham successfully withstood Nimrod, but was expelled from Babylon and left for the land of Haran (pronounced Charan, to distinguish it from Haran, Terah's son). But Abraham, the harbinger from Babylon, did not stop circulating his discovery just because he was exiled from Babylon. Maimonides' elaborate descriptions tell us, "He began to call out to the whole world, to alert them that there is one God to the whole world... He was calling out, wandering from town to town and from kingdom to kingdom, until he arrived in the land of Canaan...

"And since they [people in the places he wandered to] gathered around him and asked him about his words, he taught everyone...until he brought them back to the path of truth. Finally, tens of thousands assembled around him, and they are the people of the house of Abraham. He planted this tenet in their hearts, composed books on it, and taught his son, Isaac. And Isaac sat and taught and warned, and informed Jacob and appointed him a teacher, to sit and teach... And Jacob the Patriarch taught all his sons, and separated Levi and appointed him as the head, and had him sit and learn the way of God..."[31]

To guarantee that the truth would carry through the generations, Abraham "commanded his sons not to stop appointing appointee after appointee from among the sons of Levi, so the knowledge would not be forgotten. This continued and expanded in the children of Jacob and in those accompanying them."[32]

The astounding result of Abraham's efforts was the birth of a nation that knew the deepest laws of life, the ultimate Theory of Everything: "And a nation that knows the Creator was made in the world."[33]

Indeed, Israel is not merely a name of a people. In Hebrew, the word, Israel (*Ysrael*) consists of two words: *Yashar* (straight), and *El* (God). Israel designates a *mindset* of wanting to discover life's law, the Creator. Put differently, Israel is not a genetic ascription or attribution; it is rather the name, or direction of the desire that drove Abraham to his discoveries. Genetically, the first Israelites were mostly Babylonians, as well as members of other nations who joined Abraham's group. This was obvious to the ancient Israelites. As Maimonides wrote, they had their teachers, the Levis, and they were taught to follow life's essential laws.

Today, however, we are unaware of the fact that "Israel" refers to the desire to know life's basic law, the Creator, and not to a genetic lineage. Nearly 2,000 years of concealment of the truth, since the ruin of the Second Temple, has practically obliterated the truth that Kabbalah—the science that teaches Nature's (God's) unity—is for *all* the people in the world, just as Abraham intended it for *all* the people in Babel, and later "Began to call out to the whole world," as described by Maimonides.

Through the years, only Kabbalists kept this truth alive. Kabbalists such as Elimelech of Lizhensk,[34] Shlomo Ephraim Luntschitz,[35] Chaim ibn Attar,[36] Baruch Ashlag[37] and many other great Kabbalists wrote in plain words: *Ysrael* means *Yashar El* (Israel means straight to God).

Moreover, the need to discover this force, which we will describe in the following chapters, is as pertinent today as ever. Nothing has changed in Nature since Abraham's time, and the law of unity and oneness is still the *one* force that creates, governs, and sustains life.

In fact, today, our need to know it is more pertinent than ever because in Abraham's time, humanity had numerous roads from which to disperse, and ample land to inhabit. Today, however, we have a global community, and every crisis is on a global scale. The mistakes we make take their toll on the *whole* world. Abraham's discovery helps us add life's force into our calculations and plans, which makes it paramount, life-saving information.

The force that Abraham discovered and described to his students is the very force that drove Napoleon to conquer more than he could rule, and which is still driving China to globalize, instead of isolate. Yet, this force is also behind the voices that hail protectionism and separation. In a global world, protectionism could spell the end of our civilization. Our only hope is to unite, because unity is the direction of the force that drives all of life. Our challenge, therefore, is to learn *how* to unite. It is possible and plausible, but in a time of crisis, it will require recognizing life's force and generating a mutual effort to cooperate and collaborate, to live by this law's dictates.

The Core Desires

The importance of Abraham's discovery lies not so much in its scientific or conceptual innovation, although for his time both were absolutely radical. Rather, the primary significance of his discovery lies in its *social* aspect.

Indeed, Abraham's motivation for asking the questions that eventually led to his discovery was as much social as it was intellectual. He noticed that his townspeople were becoming increasingly alienated. For a long time, Babylonians nurtured a prosperous society that allowed multiple belief systems and teachings to coexist in harmony. But in Abraham's time, people were growing intolerant, conceited, and alienated from each other, and Abraham wondered why.

Through his questions and observation of Nature, he realized that the world that appears to our senses is but a superficial blanket that covers a complex and magnificent interaction of forces. When these forces interweave in a certain way, they induce a certain type of physical or emotional reality to appear, such as birth, death, war, peace, and all the states in between.

This interaction exists not only on a large scale, as between countries, but in every element of life, from the subatomic to the interstellar, and from the very personal to the international. In the latter part of this book I will explore the social implications of Abraham's discoveries, but to do that we need to understand more of the nature of the discoveries themselves.

Abraham's thought process in discovering these forces is evident in his questions, which to him were, as Neil Postman put it in *The End of Education*, "the principal intellectual instruments available to human beings."[38] In Maimonides' writings, Abraham asked, "How was it possible for this wheel [of reality] to always turn without a driver? Who is turning it, for it cannot turn itself?"[39] Later, his insights helped him defeat Nimrod in the debate, when Nimrod kept ordering him to serve this or that element, and Abraham kept showing him that those elements were all offshoots of something higher, without real power of their own.

Thus, through repeated pondering and observation, Abraham came to realize what really makes the world go 'round, and like all great truths, it was as simple as can be: desires, two desires, to be exact. One is a desire to give and the other, to receive. The interaction between those desires is what makes the world go 'round; it is the wheel that drives all things and the force that creates all phenomena. In Kabbalistic terminology, the desire to give is referred to as "His [the Creator's] desire to do good to His creations,"[40] and the desire to receive is described as "the desire and craving to receive delight and pleasure."[41] For short, Kabbalists refer to them as "desire to bestow" and "desire to receive."

This simple realization is what Abraham was trying to convey to his fellow Babylonians, but Nimrod tried to prevent him from doing so by trying to kill him. And when he failed to do so, he sent him away.

Alas, deporting Abraham did not restore the Babylonian spirit of camaraderie and union. Eventually, "The Lord [Creator, meaning Nature] confused the language of the whole earth; and from there the Lord scattered them abroad over the face of the whole earth" (Gen, 11:9).

This did not happen to the Babylonians because some vengeful and powerful old man called "The Lord" was holding a grudge against them. It happened to them because the desires that Abraham discovered possess a certain direction of evolution. There is no random interaction here, but a set of rules that unfold by a rigid cause-and-effect order.

When Abraham discovered these rules, he realized his local folk were headed in the wrong direction, which could only lead them to eventual destruction, so he tried his best to warn them. As we will see, these desires are as perpetual and as rigid as gravity, or the positive and negative poles of a magnet. But like gravity and the poles of the magnet, both forces can be made to work to our benefit.

To understand the similarity between the current state of humanity and the state of the Babylonian society, and hence the relevance of Abraham's discovery to the current global crises, we need to understand the direction in which the two desires evolve. And for this, we need to start from the very beginning.

GENESIS

In his book, *The Tree of Life*, the great 16th century Kabbalist Isaac Luria (the Ari), founder of Lurianic Kabbalah, today's predominant school of Kabbalah, wrote, "Know, that before the emanations were emanated and the creatures created, an Upper, Simple Light had filled the whole of reality. And there was no vacant place, such as an empty air and a void, but everything was filled with that simple, boundless Light."[42]

Since then, only one Kabbalist has ventured to compose a comprehensive explanation of these profound phrases, as well as introduced a complete commentary on *The Book of Zohar*: Kabbalist Rav Yehuda Ashlag, Baal HaSulam. In his six-volume commentary on the writings of the Ari, known as *Talmud Eser Sefirot* (*The Study of the Ten Sefirot*), Baal HaSulam explains that the Light that the Ari refers to is "All the pleasant sensations and conceptions in the world."[43] He also defines "Light" as "everything but the substance of the vessels [desire to receive]."[44]

In other words, there are only two "beings" in existence: the desire to bestow, to give, which Ashlag defines as "light," "Creator," or "pleasure," and the desire to receive pleasure, to enjoy, which he calls "a vessel," "the creature," or "the created being." To understand how the whole of reality can emerge from only two desires, we need to take a deeper look at how they interact.

FOUR STAGES AND THE ROOT OF CREATION

Electricity, gravity, and all of Nature's other forces are timeless phenomena. In other words, you cannot point to a specific point in time at which they were created because Nature's forces are not particular events; they are potentials or fields that cover the whole of space-time. They manifest under certain conditions and, given the right instruments, we can detect their existence.

To prove the existence of electricity, you need a resistor of some sort, like a lamp or a current-meter. Without something that resists the flow of the electric current, we could never know that electricity was flowing through it, and we could never discover the existence of electricity. Similarly, to prove the existence of gravity, we need to observe its effect on physical masses, and to discover light, we need an object that the light illuminates, meaning stops the light and reflects it back to our eyes.

In precisely the same way, Kabbalists discovered the desire to bestow through that desire's interaction with its resistor—their own desires to receive. When they refined and calibrated their resistors—desires to receive—they were able to detect the force that operated those desires. That was how Abraham discovered that the force that operated his desires and the rest of reality was a desire to bestow. This is the knowledge that Abraham passed on to his sons and students, and this is still the knowledge that Kabbalists pass on from teacher to student, and now to the entire world.

As a side note, the difference between one Kabbalist and another is not in the knowledge each conveys, but in the *language and style* each uses to convey it. The reason I am relying mostly on Ashlag's writings is not that he had more extensive knowledge than, say, the Ari. I am using his writings simply because he was the most recent Kabbalist, and wrote in the most contemporary style. Therefore, he is the easiest to understand for a 21st century reader with little or no background in Kabbalah. The farther we go back in time, the harder it is to grasp the full meaning of Kabbalistic texts.

Returning to the discussion at hand, in his *The Study of the Ten Sefirot*, Ashlag tells us that this desire to bestow created the desire to receive as a necessary offshoot of its wish to bestow.[45] In other words, because the desire is a desire to give, it created something that wishes to receive. Thus, just as it is impossible to explain what is a day without also understanding what is a night, or to understand the concept of "left side" without having the concept of "right side" either, it is impossible to perceive the desire to receive without perceiving the desire to give.

To put matters in the right context, when Kabbalists speak of the Creator, they are referring to the desire to give, and when they speak of Creation, they are referring to the desire to receive the Creator's giving. Also, when they are presenting a dialog between the Creator and the creatures, such as we find in the Bible, they are actually introducing a specific interaction between the desire to give and the desire to receive, not an exchange of vocalisms between a protein aggregate and a voice in the clouds.

In that regard, at the conclusion of his introduction to *The Study of the Ten Sefirot* (Item 156), Ashlag takes special care to warn us: "Yet, there is a strict condition during the engagement in this wisdom—to not materialize the matters with imaginary and corporeal issues. This is because thus they breach, 'Thou shall not make unto thee a graven image, nor any manner of likeness.' ... To rescue the readers from any materialization, I compose the book, *The Study of the Ten Sefirot* by the Ari, where I collect from the books of the Ari all the principal essays concerning the explanation of the ten *Sefirot* in as simple and easy language as I could."[46]

Thus, at the basis of existence lies not matter, but forms of desire to receive pleasure created by interactions with their Creator—the desire to give pleasure.

To tie this approach to more familiar territory, think of lightning. To the ancient Greeks, the thunderbolt was Zeus' traditional weapon. To us, the exact same thunderbolt is merely "The visible discharge of electricity that occurs when a region of a cloud acquires an excess electrical charge that is sufficient to break down the resistance of air," if we consult the Encyclopedia Britannica.[47]

Similarly, understanding the true meaning of Abraham's story requires an explanation by one who has acquired sufficient knowledge to explain it in a matter-of-fact, rational manner, meaning a Kabbalist, and preferably one of substantial understanding and sufficient didactic skills, such as Ashlag.

Going After the Thought of Creation

In "Preface to the Wisdom of Kabbalah,"[48] Baal HaSulam divides the onset of Creation into five stages and one restriction, but we can cluster them into three groups. Think of the first two groups as a car and the fuel for its engine, and imagine that the third group is the driver.

The first group contains only Stage Zero, the Root. This is the desire to give, the energy that creates and sustains the car called "Creation" (it's a very old model; they don't make them like that anymore).

The second group—Stages One and Two—builds a "platform" for evolution. This is the car itself. In a sense, the platform that the two stages have built resembles what Richard Dawkins described in *The Selfish Gene* as "The primeval soup,"[49] the oceanic substrate that contained the ingredients for life's inception.

The third group—Stages Three and Four—is "the driver." Its role is to start the engine of evolution—the interaction between the desires. As we will explain below and in the next chapter, the restriction is the wheel with which creation is driven toward its purpose: discovering the Thought of Creation.

Stages Zero and One

First, a general comment about the stages: Since Kabbalah has gained popularity in recent years, some of its terms have surfaced in various connections. The term *Sefirot* is often mentioned in relation to the origin of Creation. It is possible to describe the process of creation using the names of *Sefirot* instead of stages, but this might complicate matters needlessly. To see how the *Sefirot* and the four stages relate to the same process, refer to the essay, "Preface to the Wisdom of Kabbalah."[50]

In Kabbalistic terms, the existence of a desire to bestow without a desire to receive is called "the Root Stage" or "Stage Zero." The Root Stage is immediately followed by its mandatory offshoot—"Stage One"—the desire to receive, which is permeated with the abundance given to it by the Root, the desire to bestow.

As a result, no element in existence, from subatomic particles to the most expansive galaxies in the universe, escapes the giving-receiving "bipartisanship." It may appear in the form of hot vs. cold, dry vs. wet, small vs. big, centrifugal vs. centripetal, energy vs. matter, etc., but they all stem from the primordial opposites: giving and receiving. To portray this interaction, I use a downward arrow to denote the desire to give, and a bowl or receptacle (usually referred to as a "vessel") to denote the desire to receive (Figure no. 1).

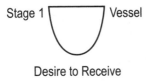

Figure no. 1: The Root Stage is immediately followed by its mandatory offshoot—"Stage One"—which is the desire to receive, permeated with the abundance given to it by the desire to bestow. The Root is known as "light" and the desire to receive, as "vessel."

Stage Two

The result of the meeting between the two desires in Stage One is Stage Two. Here is where the actual interaction between the desires truly begins. To understand the shift that occurs between Stage One and Stage Two, consider a child's admiration for its parents. Because children, especially in early childhood, idolize their parents, they strive to imitate them. They closely observe their parents' every move (with a tendency for boys to observe their fathers and for girls to observe their mothers), "study" their parents' demeanor, and try to follow suit.

Contemporary studies show how attentive children are to their parents' guidance. In *Perspectives on Imitation: From Neuroscience to Social Science*, Dr. Andrew Meltzoff and Prof. Wolfgang Prinz of Cambridge University, UK, write, "Parents provide their young with an apprenticeship in how to act as a member of their particular culture long before verbal instruction is possible. A wide range of behaviors—from tool use to social customs—are passed from one generation to another through imitative learning."[51]

Also, Dr Benjamin Spock's *Baby and Child Care* bestseller on parenting provides such a complete description of this process that I feel compelled to present it here in full: "Identification is a lot more important than just playing. It's how character is built. It depends more on what children perceive in their parents and model themselves after than on what the parents try to teach them in words. This is how children's basic ideals and attitudes are laid down—toward work, toward people, toward themselves … This is how they learn to be the kind of parents they're going to turn out to be twenty years later, as you can tell from listening to the affectionate or scolding way they care for their dolls.

"**Gender awareness.** It's at this age that a girl becomes more aware that she's female and will grow up to be a woman. So she watches her mother with special attentiveness and tends to mold herself in her mother's image: how her mother feels about her husband and the male sex in general, about women, about girl and boy children, toward work and house work. The little girl will not become an exact copy of her mother, but she will surely be influenced by her in many respects.

"A boy at this age realizes that he is on the way to becoming a man, and he therefore attempts to pattern himself predominately after his father: how his father feels toward his wife and the female sex generally, toward other men, toward his boy and girl children, toward outside work and housework."[52]

And just as a child wishes to grow up to be like its parent, Stage Two in the evolution of the desire is an expression of the wish of the desire to receive (Stage One) to be like its parent—the desire to give (the Root). This happens because as it is a desire to receive—the "offspring" of the desire to give—Stage One recognizes the Root's superiority and wishes to be like its progenitor. And because the only example that Stage One receives from the Root is that of giving, in Stage Two the desire to receive begins to want to give, as well.

Earlier, we said that at the basis of existence are forms of the desire to receive, created by interactions with their creator—the desire to give. Thus, through two natural, "automatic" reactions to giving, two opposite desires emerge: to receive (in Stage One), and to give (in Stage Two). The various combinations of these two desires form the basis of every object, every event, and every evolution that occurs in our world, including us—our bodies, our thoughts, and our actions.

Just as a child wishes to become like its role-model parent, at the root of the desire to give in Stage Two lies the desire to receive its progenitor's superior status, power, and knowledge. In other words, Stage Two is a desire to *receive* the status and nature of *giving*. For this reason, it is best to picture Stage Two as a vessel (desire to receive) that wishes to give, or "vessel of bestowal." Hence, the arrow designating this desire points outwards, toward the Creator (Figure no. 2).

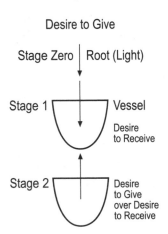

Figure no. 2: At the root of the desire to give in Stage Two lies the desire to receive. Therefore, it is best to picture Stage Two as a vessel (desire to receive) that wishes to give, or "vessel of bestowal."

But Stage Two is more than just a new desire. In wanting to give, Stage Two is admitted into an entirety new state of being. Because it no longer wishes to receive, but to bestow, it must have someone upon whom to bestow. Thus, to be like its creator—a giver—Stage Two must act positively and favorably toward others.

For this reason, Stage Two, the force that compels us to give despite our underlying desire to receive, is the force that makes life possible. Without it, parents would not have children (to whom they can bestow) or care for their offspring once they are born, life would not be possible.

Indeed, the best example of Stage Two is a mother's love for her child. If we consider the endless love, compassion, and effort mothers put into raising their babies, we are left in awe and admiration that such devotion is even possible. Yet, when you look at a mother's face while she is nursing, changing diapers, or bathing her baby, often you will find that she is glowing. Why is this so? What gives mothers the ability to not only endure such strain, but to *wish* for it and enjoy it?

The answer is simple, and every mother knows it instinctively: In giving to their babies, they experience tremendous joy. There is a desire to receive the pleasure of motherhood (or parenthood) behind every decision to bring a new life into the world. Without it, people would not have babies, unless by mistake, and this would be very unfortunate for the children.

Now we can see why Nature's initiating force is the desire to give, and not the desire to receive. Concisely capturing the essence of that concept is Baal HaSulam's Kabbalistic definition of altruism. In 1940, he published a paper titled, "The Nation." In it he writes, "The altruistic force [the desire to give] is like centrifugal waves—an outward aiming force... which flows from within outwardly."[53]

Stage Three

As Ashlag stated, the evolution of desires, which hang down by cause and effect, is mandatory, adhering to fixed, determined rules. The next mandatory step is for Stage Two to start giving, since this is what it wants to do. But in Stage Two the desire newly made desire to give has a problem to solve: it wishes to give, but all that exists besides itself (the desire to receive with its two stages) is the desire to give that created it. Therefore, the only thing that Stage Two can give to its creator is its *willingness to receive*. In other words, it will receive, just as in Stage One, but with the *intention to give* pleasure to the Root—the Creator. This "inverted" modus operandi, where the act is reception but the intention is to give, is a completely new concept and hence merits a new name—"Stage Three" (Figure no. 3)

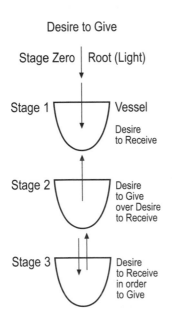

It may seem awkward to some, but we apply this mode of action routinely in our relationships. Think of a young man who comes to visit his mother after not having seeing her for a long time. It is quite likely that his mother will want to prepare something to eat for her darling son. But what if the son is not very hungry? Will he not eat? In most cases, he will eat and express his delight at the food simply because it pleases his mother.

Figure no. 3: In Stage Three, the desire to receive chooses to receive not because it enjoys it, but because this is what pleases the Root, the desire to give.

In this case, the son is not focused on his own pleasure, but on his mother's pleasure in watching him eat. In "Preface to the

Wisdom of Kabbalah,"[54] Baal HaSulam describes this mode of work as partial use of the desire to receive, just the necessary minimum for the reception of pleasure, while maintaining the center of attention on the giver's delight at the receiver's acceptance. In our culinary example, the son must have *some* appetite or he will not be able to eat at all. However, his appetite should not be big enough to shift his intention (or attention) from pleasing his mother to pleasing himself.

Stage Four

When the son's appetite is mild enough to be subordinate to his desire to please his mother, he can focus on his intention to please, rather than on his stomach. But what if he were very hungry and had not eaten for a whole day? Would he still be able to ignore his rumbling stomach, focus only on his mother's pleasure, and eat only because it pleases her? When Stage Three begins to receive because it wishes to please the Root, it realizes that the more it receives, the more it pleases its maker, the Root.

In consequence, it begins to wish to receive more and more and more. Finally, it wishes to receive *everything*, just as in Stage One, thus reawakening the whole of its desire to receive. This *self-evoked* total desire to receive is called "Stage Four."

Yet, there is a fundamental difference between Stage One and Stage Four: relation to the giver. Stage One does not relate to the giver, only of the abundance. As soon as it "realizes" there is the desire to give that created it, it wishes to be like the giver, and this initiates Stage Two. Stage Four is realizes not only the giver's existence, but also of the giver's benevolence and *primacy*, since it is the desire to give that initiated creation. And being a complete desire to receive, Stage Four wishes to receive not just the abundance that Stage One receives, but the *primacy status* of the Root (Figure no. 4).

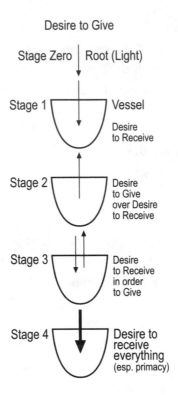

Figure no. 4: Being a total desire to receive, Stage Four wishes to receive not just the abundance that Stage One receives, but the *primacy status* of the Root.

However, to receive such a status, Stage Four must be Creator-like, and it is not. Instead, it is a conscious desire to receive everything—omnipotence, omniscience, and even the *nature* of the Creator. Anything less than that would be incomplete, since it would not be precisely identical to the Creator. This is what Ashlag means when he writes in "Preface to the Wisdom of Kabbalah"[55] that Stage Four wishes to achieve the Thought of Creation (Figure no. 5).

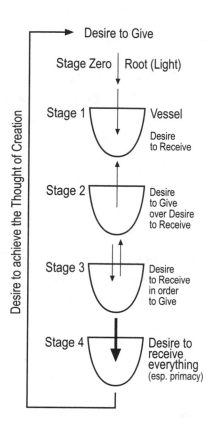

Figure no. 5: Stage Four wishes to achieve the
Thought of Creation

In another essay, "The Giving of the Torah [Light]," Ashlag offers a beautiful explanation of the nature of the Creator-created relationship that occurs at the outset of Creation: "This matter is like a rich man who took a man from the market and showered him with gold, silver, and all desirables each day. And each day he gave him more gifts than the day before. Finally, the rich man asked, 'Do tell me, have all your wishes been fulfilled?' He [the man] replied, 'Not all of my wishes have been fulfilled, for how good and how pleasant it would be if all those possessions and precious things came to me through my own work, as they have come to you. Then I would not be receiving the charity of your hand.' Then the rich man told him, 'In this case, there has never been born a person who could fulfill your wishes.'"[56]

This resentment of gifts was well observed in research conducted by Amani El-Alayli and Lawrence A. Messe of Michigan State University. Their findings, published in the *Journal of Experimental Social Psychology*, were that when receiving unexpected favors, people may experience two opposing emotions: a desire to reciprocate the favor, which the researchers correctly described as "obligation," or resentment, which they referred to as "psychological reactance."[57]

Moreover, they wrote, "Participants' evaluations of the supervisor [benefactor] suggested that people have mixed impressions of someone whose favors violate [exceed] expectancies or norms."[58] This research clearly demonstrates that it is a natural human trait to feel shame and embarrassment when treated with exceptional generosity. These emotions, Kabbalah explains, are directly rooted in the shame that Stage Four experiences when faced with unbounded giving without the chance of becoming a giver, too.

Thus, when Stage Four realizes it cannot obtain the Root's primacy, it realizes that it cannot receive *everything* and that it is inherently inferior to its maker. This instantly extinguishes any sensation of pleasure in Stage Four, and despite the infinite abundance that the Root gives, Stage Four remains with a sense of emptiness, since its biggest wish has not been granted. In Kabbalah, when Stage Four's desire to be like its maker overshadows all other pleasures, it is called "restriction." Because the desire to be like the Creator is so much greater than all other desires, it practically prevents pleasure from being experienced.

From here on, evolution will unfold in a single underlying purpose: to repossess that great abundance that the Root wishes to give, and which can be received only with the intention to bestow.

3
HUMANITY'S SHARED ORIGIN

In the previous chapter, we talked about the emergence of the desire to receive in Stage One, and the desire to give in Stage Two, as offshoots of the primordial desire to give in the Root. We also showed how because of its desire to give, the desire to receive was reactivated in Stage Three and maximized in Stage Four. Maximizing the desire to receive caused it to want not merely to enjoy, but to actually become like its progenitor—the Root Stage—and even to have the status of the Root Stage's primacy. The subsequent realization that this was not (yet) possible induced a sense of inherent inferiority in Stage Four, which induced a restriction—elimination of any sensation of pleasure (light).

Also, because Stage Four's real desire is for the Root's primacy, it does not settle for the unbounded pleasure received in Stage One. Instead, it wishes to obtain the *nature* of the Root, the Thought of Creation, and consequently the Root's primacy.

Thus, the elimination of pleasure in Stage Four is neither a result of its inability to receive, nor a consequence of the Root's inability to give. The Root gives incessantly, but the desire to receive does not *want* to receive something as degrading as charity (as described by Ashlag in "The Giving of the Torah"[59]). For this reason, because Stage Four wishes to acquire the giver's thought and become like its Creator, its restriction is an offshoot of its decision to not receive unless with the intention to bestow, as this reciprocates the Creator's desire to bestow.

To achieve that, Stage Four builds a three-part mechanism, called *Partzuf* (Face), to determine if it should receive light, and if so, how much, with the aim to bestow at any given moment (Figure no. 6). The top section of the *Partzuf* is called *Rosh* (Head). Its task is to determine how much of the abundance (light) is to be received by the desire to receive. The desire to receive itself constitutes the bottom part of the *Partzuf*, which is called *Guf* (Body).

Between the *Rosh* (Head) and the *Guf* (Body) stands the *Masach* (Screen). Just as a selectively permeable membrane allows only some molecules to seep through it, the *Masach* screens out the light, allowing into the *Guf* only as much light as the *Rosh* decided it could receive with the intention to bestow, while repelling the remainder of the light. That way, the *Masach* functions as a guard, making sure the degradation sensed immediately prior to the restriction will not return.

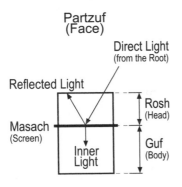

Figure no. 6:
The mechanism called *Partzuf* (face): The *Rosh* (head) determines how much of the abundance (light) to receive. The *Guf* (body) is desire to receive itself, and between the *Rosh* and the *Guf* stands the *Masach* (screen), which admits into the *Guf* only as much light as can be received in order to bestow.

In a sense, a *Partzuf* can be compared to a big company, where the *Masach* is like the Human Resource (HR) department in that company. If the management, the *Rosh* (head), wishes to increase production (bestowal, level of being like the Creator), it needs to hire more people (desires) so it can receive more light/ pleasure (thus bestowing upon the giver). Once new people are hired, they will be admitted into the company (*Guf*, body) and will be put to work: receiving pleasure in order to bestow.

When the *Rosh* decides it is time to act, the *Masach*—HR department—screens the applicants (desires) and chooses only the right ones. A new employee (desire) must not be under-qualified (too small), since that would not bring pleasure to the Creator (because you cannot experience great pleasure when you have a small desire for it). It also cannot be over-qualified (desires that are too intense to be used with the aim to bestow), since that would reawaken the excessive desire to receive and the creature's resulting degradation.

Yet, in both the *Partzuf* and its mundane "work-alike" company, which we can call "Creation," a problem remains unsolved: What about the desires (people) who were not hired (made to work in order to bestow in the *Guf* of the *Partzuf*)? Are they doomed to eternal unemployment (rejection)? That would mean that there will always be lights (pleasures) that the Creator wishes to impart, but which we cannot receive. This defies the purpose of creation: for the recipients (Creation, us) to receive unbounded delight, the power, knowledge, and supremacy of the Creator.

Indeed, eventually all the desires will be "hired" and put to work, and all the lights will be received. However, to avoid overloading the system and risking a total collapse, some desires must be temporarily put on hold. The lights that should be received in those desires are therefore reflected and remain as "surrounding lights" (Figure no. 7).

The desires and lights that cannot be put to work for the time being apply constant pressure on the *Partzuf*, "reminding" it that there is still more pleasure to receive if it is to receive from the Creator everything that the Creator wishes to impart. In our mundane example, the marketing department is the "surrounding light"—constantly reporting new potential markets into which the company can expand, and which can produce substantial profits.

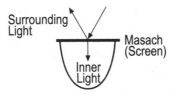

Figure no. 7:
While the *Partzuf* is incapable of receiving all the light, the reflected light must remain outside the *Partzuf*. This is called "surrounding light."

HOW DESIRES BECOME WORLDS

Continuing the *Partzuf*/company allegory, the company, a.k.a. "Creation," begins to sort out the "unemployed" desires on its waiting list, placing the weakest, easiest to handle desires at the top of the list, and the most intense, unruly ones at the bottom. Creation divides these desires into four categories, similar to the four stages in the evolution of desires. It refers to each category as an *Olam* (world), from the Hebrew word *Haalama* (concealment), since these desires must be kept separated and concealed from the lights until they can be operated correctly—with the aim to bestow. Thus, the desires with qualities most similar to Stage One are called "the world of *Atzilut*," those most similar to Stage Two form "the world of *Beria*," with those most similar to Stage Three forming "the world of *Yetzira*," and those most similar to Stage Four becoming "the world of *Assiya*" (Figure no. 8). For short, they are called "*ABYA*."

Four Worlds Akin to Four Stages

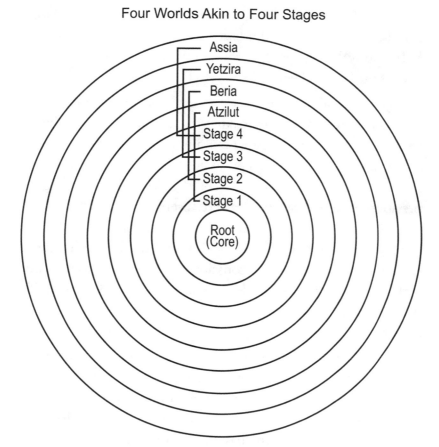

Figure no. 8: Creation divides the remaining desires into four categories, akin to the four stages of the evolution of desires. It refers to each category as an *Olam* (world), from the Hebrew word *Haalama* (concealment).

When Kabbalists describe the spiritual realm—where desires work with the aim to bestow—they usually divide it into worlds and describe what happens in them (how desires actually receive). Therefore, they often refer to everything that precedes the worlds of *ABYA* as a world as well, and call it "the world of *AK*" (*Adam Kadmon*—the primordial man). In a way, the world of *AK* parallels the Root Stage, or Stage Zero, in the evolution of desires.

Note that our world is not mentioned among the spiritual worlds. Because our world is based on egoism, and the worlds in Kabbalah reflect levels of bestowal, our world is not considered part of the spiritual (with the aim to bestow) system.

The spiritual system is ceaselessly evolving through interaction among its forces, gradually making more of its desire capable of receiving with the aim to bestow, building each stage on conclusions and actions performed in its preceding stages. Similarly, as a baby grows, its physical and cognitive abilities develop by building on previously acquired capabilities and observations. Without going through these early stages of development, babies would not become adults. Of course, we do not, and need not keep those early-life observations in our consciousness while we go about our daily routine, since they have become automatic; yet, we constantly use them in our lives as adults.

With children, we help them acquire new capabilities and data, and we watch over them to make certain that they do not attempt to do things prematurely. Similarly, to complete creation's "maturing" into being Creator-like, it needs to learn with which of its desires it can work (receive in order to bestow), as well as how and with which desires it still cannot work because it would reawaken the sense of inferiority and shame.

Hence, in each world, Creation carefully examines the light (pleasure) that the desire to give wished to impart on it. In *Atzilut*, Creation receives all the light, since *Atzilut* corresponds to the desire of Stage One—receiving all the light "automatically," not getting its own desire to receive involved. For this reason, the desire-pleasure combination in *Atzilut* is called "still" or "inanimate," since the desire is passive, still.

In *Beria*, creation receives less light because *Beria* corresponds to Stage Two, which is a more developed state of the desire to receive—a desire to give, like the Creator. Because *Beria* corresponds to the first desire that reacted to the light, it was given the name of the first level of life: "vegetative."

In *Yetzira*, Creation receives even less light than in *Beria* because *Yetzira* corresponds to Stage Three in the desire to receive, which received only a little bit of light to begin with (look back to Chapter 2, section "Four Stages and the Root of Creation"). Still, it is a more developed stage in the evolution of the desire to receive, showing a certain measure of autonomy. For this reason, it received the name of the level in evolution whose members display at least some autonomy—"animate."

In *Assiya*, Creation receives so little of the light that it is not sensed as pleasure whatsoever, but as mere sustenance. *Assiya* corresponds to Stage Four in the evolution of desires, and just as Stage Four experienced the restriction, the world of *Assiya* is barred from experiencing the light. But because it corresponds to the latest, most developed, and most complex level of desire, it received the name of its corporeal parallel: "human" or "speaking."

PARALLEL NAMES

In "The Essence of the Wisdom of Kabbalah," Baal HaSulam explains that the worlds *ABYA* are all very similar to one another: "Kabbalists have found that the form of the four worlds, named *Atzilut*, *Beria*, *Yetzira*, and *Assiya*, beginning with the first, highest world, called *Atzilut*, and ending in this corporeal, tangible world, called *Assiya*, is exactly the same... This means that everything that eventuates and occurs in the first world is found unchanged in the next world, below it, too. It is likewise in all the worlds that follow it, down to this tangible world.

"There is no difference between them, but only a different degree, perceived in the substance of the elements of reality in each world. The substance of the elements of reality in the first, Uppermost world, is purer [more giving], than in all the ones below it, and the substance of the elements of reality in the second world is coarser [more receiving] than in that of the first world, but purer than all that is of a lower degree.

"This continues similarly down to this world before us, whose substance of the elements in reality is coarser and darker than in all the worlds preceding it [most receiving, to the point of egoism]. However, the shapes and the elements of reality and all their occurrences come unchanged and equal in every world, both in quantity and quality."[60]

Therefore, although Kabbalah speaks of desires, not of physical objects, because all worlds are practically identical, Kabbalists often use names of objects or processes from the physical world to explain spiritual states or processes that occur on the level of desires. Physical examples are much clearer and more tangible. The term *Partzuf* (face), which we discussed above is one such case. A "sassier" example would be *Zivug de Hakaa* (coupling by striking), which is a code name for describing the entire process of repelling light (striking) and then receiving (coupling) only the amount of light that can be received in order to bestow.

Accordingly, in his "Introduction to the Book of Zohar," Ashlag explains that the name, "still," was given to the world of *Atzilut* because it consists of the desire to receive in Stage One, which is completely passive.[61]

The corporeal equivalent of the world of *Atzilut* is minerals. All minerals strive (wish) to maintain their form. They have no desire to become anything other than what they already are; if you try to change them into something else, you will have to apply energy and manipulation on them because they will resist the change.

In Ashlag's words, "Phase One of the will to receive, called 'still,' ...is the initial manifestation of the will to receive in this corporeal world. ... But no motion is apparent in its particular items. ... And since there is only a small will to receive... its power over the particular items [minerals] is indistinguishable."[62]

Beria received the name, "vegetative," since it is the beginning of an independent desire. As might be expected, the material manifestation of this desire is plants. Plants grow, blossom, and shrivel, and each plant is a distinct entity, as opposed to the aggregate of molecules that forms minerals. Yet, plants have no free choice in their movements. When plants of a certain kind grow in close proximity, they will all behave in exactly the same way. For example, the head of a sunflower plant will always turn toward the sun (Image no. 1), and all wheat stalks turn yellow when harvest time approaches.

Image no. 1: the head of a sunflower plant will
always turn toward the sun.

Yetzira was named, "animate," and corresponds to Stage Three of the desire to receive. In *Yetzira*, Creation enjoys a substantial measure of "freedom and individuality ...a unique life for each item," writes Ashlag in the above-mentioned introduction. Yet, in *Yetzira*, he explains, "the desire still lacks the sensation of others, meaning there is no preparation to participate in others' pains or joys."[63]

Assiya was named "speaking" or "human," as it reflects the complete and most complex form of the desire to receive. At the human level, and Ashlag explains that this is a fundamental difference between the speaking and the animate levels, the will to receive includes the sensation of others[64]: "The will to receive in the animate, which lacks the sensation of others, can only generate needs and desires to the extent that they are imprinted in that creature alone. But the human, who can feel others, too, becomes needy of everything that others have, and is thus filled with envy to acquire everything that others have." For this reason, "when one has a hundred, he wishes to have two hundred, and so his needs forever multiply until he wishes to devour all that there is in the entire world."[65]

To truly understand the difference between the human level of desires and all other levels, consider the following experiment: Offer a dog a new touch-screen smartphone instead of its favorite dog food and see which of them it chooses. Afterwards, replace the dog food with human food and leave in the smartphone. Then, try the same experiment with a person.

THE BIRTH AND FALL OF ADAM

Thus far, we have discussed the origin of Creation. We explained how Creation receives what pleasure it can in order to bestow, and builds itself to be as similar as possible to its Creator. However, even after all the worlds have been established in the *Partzuf* (company), and all the lights that could be received in order to bestow are received in the *Partzuf*, there still remains one desire that cannot be made to work in the *Partzuf*—the desire to be like the Creator. This is the desire that the host in Ashlag's allegory was referring to when he said (Chapter 2), "In this case, there has never been born a person who could fulfill your wishes."[66] This is the most intense desire, the core desire of Stage Four, and at the same time it is utterly unachievable.

So once all the desires were exploited to the maximum, Creation's (the company) marketing department (surrounding light), reminded the company management—the *Rosh* (head) of Creation—that there was still more light to be received. Now it was the *Rosh's* duty to examine this new desire and determine if it could receive this desire with the intention to bestow.

For this reason, the *Rosh* assembled a special board meeting to discuss the fate of this last desire. In this meeting, the argument for not using it was that it was too strong to handle. Indeed, how can one handle a desire to be like one's parent? If the *Partzuf* actually received what it wished for in that desire, it would be similar to a child instantly becoming an adult, without the knowledge and experience acquired over the years of growing up. Clearly, this was too complicated and too dangerous a desire to handle.

"On the other hand," argued other directors, "If we consider the nature of this desire, we will realize that there cannot be any danger in it. In fact," they claimed, "it is fail-safe."

"How so?" wondered the opposers. "It is fail-safe because of the nature of the desire itself—to be like the Creator, a giver. How dangerous can it be to want to give?"

The advocates convinced the opposers, and the decision was made for Creation to hire the biggest desire—the wish to be like the Creator. To do this, Creation built a distinct *Partzuf*, called *Adam ha Rishon* (The First Man), and assigned it the task of operating and managing the last and greatest desire of all.

However, the decision to try to receive the last and greatest of all pleasures turned out to be a fatal error. What Creation did not know was that the biggest light, which comes with the biggest desire, has a gift attached to it. When you become Creator-like, you become Creator-like in the full sense of the word, not just in your *desire* to give, but also in your *ability* to give—to *create*—you become omnipotent and omniscient. This was a pleasure Creation could not receive with the intention to bestow.

As soon as Adam, the specially designed *Partzuf*, began to receive the light, he (Adam) discovered the gifts attached to the light, and they were so blindingly enticing that he completely forgot about the intention to give.

And the minute Adam began to think in this way, he tried to act on it, to be a creator. However, to create you need a desire to give, and Adam did not have it. This reawakened the inferiority and shame that were covered by the initial *Masach* in Stage Four, and with it, the light vanished, just as it did during the restriction.

But Adam's desire could no longer be reversed; he saw what pleasures await those who become like the Creator and could not forget it. And for this reason, Adam could not be made to work in order to bestow, since he knew that if only he could find a way to be like the Creator, he would be the sole ruler of the universe, of the whole of reality. Thus, Adam became selfish to the core, each part of him wishing to be like the Creator. And in consequence, the selfish parts disintegrated into myriad fractions, each with its own selfish little desire to become like the Creator.

The disassembling of Adam's *Partzuf* is known as "the breaking of Adam's soul" or "the breaking of the soul" for short. With Adam's shattering, a new entity appeared in reality—an egoistic entity—whose desire is to bestow upon itself, rather than upon the Creator, and whose ultimate wish is for omnipotence and omniscience, rather than for total bestowal.

In Kabbalah, explains Baal HaSulam in the "Preface to the Wisdom of Kabbalah," the difference between spirituality and corporeality is that in the spiritual realm, there is no desire to receive without a *Masach*, while in the corporeal realm there is only a desire to receive without a *Masach*.[67] Hence, our universe is the only corporeal realm in existence, and all that exist in our universe are the offspring of the shattering of Adam's soul.

The reason why we consider our universe a "world," the same term we ascribe to the spiritual worlds, is that a "world" reflects a certain measure of concealment of light. The only difference between our corporeal universe and the spiritual worlds is that in a spiritual world, even when there is no light at all, there still awareness of the Creator's quality of bestowal and there is a desire to have it. In our universe, there is such complete concealment that we are not even aware of the meaning of the word, "Creator," and think of it as an entity (if not a person) that awaits our pleas in return for a merciful response.

In Hebrew, humans are called *Bnei Adam* (the children of Adam). Indeed, we are the offspring of Adam's mistake, and it is therefore only we who can mend his mistake. Being the only species that can choose its course in life, humans are the only ones who can determine the fate of all life on earth—for better or for worse.

As we will see in the following chapters, save for humans, the whole of Nature abides by a rule that aligns it with the laws of the spiritual worlds. We, on the other hand, must learn to abide by that rule by ourselves. By wanting to have the intention to give more than the gift that comes with giving (omnipotence and omniscience), we can mend Adam's mistake. That is, by choosing the intention to give, the gift will still be attached to it and we will still receive omnipotence and omniscience. However, because we will have the intention to give, we will receive the gift of being Creator-like because we will know that by doing so we are pleasing the Creator, who wants to give us this gift. As a result, we will enjoy the gift, but will not be broken—falling into self-centeredness—as it happened the first time. This will be the end of correction for the whole of humanity, and attainment of the purpose of Creation, as intended in the Creator's thought of Creation.

In the next chapter, we will explore how life evolved in the corporeal (physical) world after Adam's shattering, what parts of Creation have already been corrected, and what still awaits our correction: choosing to give rather than to receive.

4

THE UNIVERSE AND LIFE ON EARTH

At the end of the previous chapter, we said that Adam's shattered soul is our common origin. Being a *Partzuf*, Adam's structure was a perfect replica of its parent (corrected) *Partzuf*. In breaking, Adam extended the structure of the spiritual worlds (worlds of bestowal) to its lowest point—ultimate reception.

In consequence, all that exists in the spiritual worlds exists in our world, as well. For this reason, the same four-stage pattern by which the stages of desire evolved, followed by the four-stage evolution of the spiritual worlds, exists in our physical world. As we explore how our world has evolved, we should keep in mind the desires that evoke and guide it.

A BIG BANG

Time, as we know it, began approximately fourteen billion years ago. From the Kabbalistic, spiritual perspective, the "big bang" was the shattering of Adam's soul. The reason we see it as a material event is that we see the world through corporeal (self-centered) eyes. If we could see it from the perspective of the force that induced this massive explosion we call "the big bang," we would see it as an outcome of Adam's attempt to receive using the last, and greatest desire, as described in the previous chapter.

THE FOUR STAGES IN MATTER

As the original desires evolved in stages, their mundane parallels appeared and corrected one at a time, from the easiest to the hardest. Now, as each desire manifests itself in our universe, Nature, which we said in Chapter 1 is synonymous with the Creator, must "teach" it to work so that it contributes to the well-being and sustainability of the universe.

To accomplish this, Nature applies a very similar approach to Darwin's natural selection principle. In fact, many leading scholars now acknowledge the existence of the natural selection process in the period before the advent of life on earth. Professor Ada Yonath, Nobel Prize laureate in Chemistry, made the following statement in an international convention celebrating the 150th anniversary of the publication of Darwin's *On the Origin of the Species*: "The survival of the fittest and natural selection played an important role in the pre-biotic world, even though these qualities are related primarily to the evolution of the species."[68]

As in Darwin's natural selection principle, the merit of any new development in Nature is judged by its contribution to the sustainability of its beneficiary. The difference between the Darwinian principle and the Kabbalistic one is the beneficiary: in Darwin's classic theory, the beneficiary is the species; in Kabbalah, the beneficiary is Nature--the *whole* of Nature, meaning the Creator.

If this concept sounds a bit far-fetched, think of a species as part of its ecosystem. In contemporary biology, it is common to view a species in relation to its surroundings, rather than irrespective of it. And since we now know that all ecosystems are connected, it is easy to understand that a disturbance in one system can and will adversely affect the rest of the systems on the planet.

Perhaps the best description I have heard to date, explaining how Nature shifts its elements from receiving from their environment to giving to their environment, came from evolutionary biologist Elisabet Sahtouris, PhD. In a presentation she gave in November 2005, at a conference in Tokyo, Dr. Sahtouris stated, "In your body, every molecule, every cell, every organ and the whole body, has self-interest. When every level... shows its self-interest, it forces negotiations among the levels. This is the secret of Nature. Every moment in your body, these negotiations drive your system to harmony."

Clearly, the balance and well-being of all systems is imperative for the survival of the human body. As a result, balance is just as imperative for the survival of each of the body's systems. Today, the view of Nature as a system rather than a collection of separate elements has gained ground among leading researchers. It has led to the emergence of such fields of science as ecology, cybernetics, systems theory, and complexity.

As we have already seen, Kabbalah has always regarded the whole of Nature as a single unit. This wholeness applies not only to earth and to life upon it, but to the entire universe— the corporeal part of it, as well as the spiritual one.

Hence, the same rules that apply to the spiritual world— the world of altruism—apply to our corporeal world—the world of egoism. The difference between our world and the spiritual one is that spiritual desires are all about bestowal, while we are descendants of Adam's shattering. As such, we are inherently self-centered, at times to the point of obliviousness even to this fact that we are so.

And because we are so self-absorbed, we are unaware of the fact that at its deepest levels, Nature is governed by altruistic rules. The role of Kabbalah is to uncover these rules and introduce them as a way to understand our world and manage it on a new level of awareness. For this reason, everything I will discuss henceforth, from the formation of the universe to the mending of human relations, will derive from and rely on the concept of evolution of desires I have explained thus far.

Still

Following Adam's shattering, each piece in the desire to receive begins to feel like an independent self, separated from its environment and wishing to absorb from it. This desire to absorb, the pulling force, or gravity—the physical parallel to the desire to receive—caused the first clusters to form in the universe, which later became the substance of the first galaxies in the universe.

As space and gravity fields created more structured forms of the desire to absorb (meaning the desire to receive), particles appeared. The absorption process continued and stars were born with planets surrounding some of them. Thus, gravity, the weakest force in Nature, created the infrastructure of the entire universe, just as Stage One, the weakest desire to receive, created the infrastructure for the Four Stages and all the spiritual worlds that followed.

As in Stage One, the desire to receive in the corporeal inanimate consists primarily of a wish to secure its own persistence, to sustain itself. Its only relation to others is that it resists any attempt to break, dissolve, or otherwise change it. Yet, as a result of the inanimate level's aspiration to maintain its own persistence, some particles "discovered" that they could best secure their future by collaborating with other elements.

Unlike Darwin's theory of evolution, Kabbalah asserts that there is no coincidence. Particles do not really "discover" or happen to collaborate and subsequently benefit from doing so. This would imply that Nature is purposeless, random, that there is no predetermined goal at the end of the process. Instead, Baal HaSulam explains (in "Preface to the Wisdom of Kabbalah,"[69] *The Study of the Ten Sefirot*,[70] and in other places) that since our world is the last in a series of cause-and-effect events, the desires that appear in our world already contain (albeit not consciously) recollections of previous states within them, since they are their offshoots. Hence, the desire to receive in this world already has a recollection of the Four Stages, the *Partzuf*, and all the spiritual worlds. As a result, the preparation, the set-up for discovering the benefits in collaboration, pre-exist in all the levels of desires in this world. This is what allows them to "miraculously" discover the benefits of "negotiating into harmony," as Sahtouris put it.

Most physicists agree that particles did not need much time to "discover" the benefits of collaboration. A publication by the Haystack Observatory, a research center at MIT, explains, "When the universe was 3 minutes old, it had cooled enough for these protons and neutrons to combine into nuclei."[71] However, to develop further, they had to forge additional collaborations, which manifested in the form of electrons. These balanced the positive charge of the nuclei. This is how the first atoms appeared.

To those particles, being part of an atom—and thus yielding their own interest in favor of the interest of the atom—was all the correction they needed. In acting for the good of the system they lived in instead of for their own good, they stopped being self-centered and became system-centered. They were now "aware" of their environment and how they could contribute to it. In doing so, they became "altruists," albeit for the selfish reason of perpetuating their own existence.

The "reward" for particles that excel in giving to their environment is the creation of a strong environment, meaning stable atoms. This guarantees their future existence.

Moreover, because atoms need all their particles to maintain themselves, the atoms themselves protect the particles within them. Thus, by yielding their self-interest in favor of the interest of their atoms, particles gain the entire system's interest in the well being of these atoms. This "deal" proved so successful that "Moments after the Big Bang, protons and neutrons began to combine to form helium-3 and other basic elements," said Robert Rood of the University of Virginia, as quoted in a release by the National Radio Astronomy Observatory.[72] Thus, the first minerals emerged.

The human body is possibly the most vivid example of the *modus operandi* of yielding self-interest before the interest of the host system in return for the system's protection. In the human body, as in any organism, each cell has a particular role. For the organism to persist, each cell must perform its function to the best of its ability and replace the goal of maintaining its own life with the goal of maintaining the life of its host organism. If a cell begins to act contrary to that principle, its interests will soon clash with those of the body and the body's defense mechanisms will destroy it. Otherwise, it is likely to create a tumor of insubordinate cells that strive to consume the body's resources for their own benefit. When such a process occurs, we diagnose it as "cancer."

If the cancer wins, the body dies and the tumor dies along with it. If the body wins and the cancer dies, the body persists, along with the cells of the organ that did not become malignant, and the self-centered cells are extinguished. This is Nature's failsafe mechanism for ensuring that self-centered systems will not exist. Here, too, there is nothing miraculous; it is simply that self-centered mechanisms invariably consume themselves to extinction because they end up consuming their food supply.

Thus, it is in the interest of all cells in the body to dispose of the tumor. Put differently, to guarantee the survival of elements in a system, the elements in that system must cater to the well-being of the system *before* they cater to their own well-being. In return, the system will cater to their well-being and provide for their survival.

The principle explained just now is valid not only for particles, atoms, and organisms, but for all of life. By applying it, all elements in Nature learn to yield their self-centered natures to an altruistic nature, which considers the good of the collective before its own good.

Thus, returning to our topic of observing the early universe, once particles joined to create atoms, atoms began to bond, thus creating the first molecules. These adhered to the same rule, and the molecules that survived were those in which the atoms were tightly connected, just as with atoms, yielding their self-interest before the interest of their host systems—the molecules.

In this entire process, there is no freedom of choice. An atom or a molecule cannot choose to *not* be created, since the elements that constitute it find it in their best interest to form it in order to best protect their interests. Yet, by creating molecules, atoms accomplished something far more significant than protecting themselves and the particles that created them. Like particles, they built a system before which they could yield their self-interest, and by so doing, atoms transformed from being self-oriented to being system-oriented, meaning altruistic.

In this way, another layer of the inanimate level of desire to receive was corrected. And although there was no free choice in this correction, the altruistic *modus operandi* is all that is required of minerals to be considered corrected. As Stage One did not have any free choice in its evolution, the inanimate has no free choice in its evolution; it simply works to ensure its persistence as best it can.

Interestingly, Darwin's theory reflects almost the same pattern in its principle of natural selection. One difference between Kabbalah and Darwinism is that what Darwinism defines as stable molecules vs. unstable ones, Kabbalah defines as balanced molecules vs. unbalanced ones. Balanced molecules support the atoms that comprise them, and the atoms equally support their moleIn The Selfish Gene, Richard Dawkins—one of Darwin's most renowned contemporary proponents—describes the process of molecular evolution: "The earliest form of natural selection was simply a selection of stable forms and a rejection of unstable ones. There is no mystery about this. It had to happen by definition."[73]

Dawkins' observations are congruent with those of Kabbalah. In Kabbalistic terminology, a stable molecule is one whose atoms have yielded their self-interests in favor of the interests of the molecule. Thus, Dawkins' "stable forms" are synonymous with Kabbalah's "corrected molecules," in which atoms have become "altruistic." Conversely, in unstable (uncorrected) molecules, one or more of the atoms remained focused on its own interest.

Following the same procedure as particles and atoms, molecules began to congregate and create what biologists refer to as "molecular interactions," or "bonds." As with molecules, interactions in which molecules dedicated themselves to the strength and well-being of the bond prospered, and those whose molecules were only partially supportive of their bond disintegrated.

Many forms of molecular interactions exist in nature, but less than four billion years ago, one particular interaction marked the shift between the inanimate stage on earth (and perhaps in the universe) and the vegetative one. This special aggregate of molecules was given the name, "Deoxyribonucleic acid," otherwise known as DNA (Image no. 2).

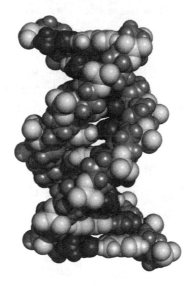

Image no. 2:
Deoxyribonucleic acid, also known as DNA

In atoms, particles assume different roles: some form the nucleus and some form the shell, for example. Similarly, in molecules, atoms assume different roles and must adhere to rigid forms of connections. And finally, in molecular interaction, each molecule plays a different role.

But with the appearance of DNA, things began to change. DNA is not yet another structure made of different molecules that form a structure. It is a structure that can *interact* with other structures, where each structure is assigned a different function. These, combined, serve the good of the *structure*. In biology, these structures are called "cells" or "unicellular organisms" and they constitute the most primitive form of life.

You could argue that essentially, these organisms function much the same as atoms, molecules, or molecular structures introduced before. But the unique structure created around the DNA allows for two hitherto nonexistent functions to occur: 1) DNA is the first known structure in nature that can replicate itself, as well as the molecular structures that support it! 2) Cells are the first structures that systematically interact with their environment. They absorb nutrients from their environment, process them to extract the energy they need for survival, and then secrete the waste. Moreover, cells can repeat this process accurately so many times that they can actually *change their environment*.

There are many definitions of life. To be on the safe side, I will choose the one that *Encyclopedia Britannica* offers: "Matter that shows certain attributes that include responsiveness, growth, metabolism, energy transformation, and reproduction."[74] The first cells, named "prokaryotes," had all those attributes and were a direct evolution of molecular interaction. Thus, the beginning of life as we know it was prompted by the same law by which all systems achieve balance and sustainability—their constituents yielding their self-interests before the interest of their host systems, in return for the system's care for them

Vegetative

As we said above, the first living organisms were primitive cells, known as "prokaryotes." As with minerals in the inanimate phase, prokaryotes grew more complex.

The vegetative phase in the evolution of life corresponds to Stage Two. The difference between Stage One and Stage Two is that Stage One is passive—receiving what Nature gives it—while Stage Two reacts to it, wishing to give back. Similarly, plants respond to their environment and interact with it. Their product, oxygen, is the gift of the flora to our world and is such a vital element of life that without it, evolution as we know it would not be possible.

In his "Introduction to the Book of Zohar,"[75] Ashlag explains that the vegetative level of the desire to receive, as manifested in plants, exhibits a more intense desire to receive. This is why the structures it creates are more complex and have a more noticeable impact on their environment.

Also, unlike minerals, plants are individual specimens with their own reproduction, feeding, and even migration mechanisms. Yet, like minerals, all plants behave similarly—accurately adhering to the program installed within them by the Creator. They open their petals (if they have them) at the same time in the morning, close them at the same time in the evening, and follow almost exactly the same procedure as do the other specimens in their species.

Thus, compliant with the law of yielding self-interest described in the previous section, cells continued to evolve, producing increasingly intricate and complex structures. At first, they congregated in large colonies of single cells. Then, gradually, they began to realize that they could benefit from ascribing different roles to different groups of cells. Some cells became "hunters," providing food for the entire colony, other cells became guards, others still became cleaners, and each group contributed its best to the community.

As we said earlier about the collaboration of particles, collaboration of differentiated organs is not coincidental. It relies on similar structures that exist in the spiritual, altruistic realm. The description of the spiritual (altruistic) worlds we provided in Chapters 2 and 3 is a very basic depiction of them. In *The Study of the Ten Sefirot*,[76] Baal HaSulam provides a detailed examination of the internal structure of the *Partzuf* we discussed earlier, and explains about such systems as the digestive system, the reproduction system, hands, legs, etc.

However, Baal HaSulam describes all these elements as interactions between desires to bestow and desires to receive. These are *not* physical objects of any kind, although how they behave serves as a "prototype" for the behavior of similar systems in our world. In Kabbalah, a prototype is called "root" and all its offshoots are called "branches."

Beyond the obvious advantage of size that colonies have over single cells, returning to the topic of evolution, cells in colonies have another edge over single cells: they can focus on a single task and thus perfect their performance, increasing their contribution to the colony and relying on their fellow cells in the colony to provide for their other needs.

Single cells, on the other hand, had to perform all the necessities of sustenance by themselves. This heightened efficiency meant that colonies spent less energy to produce the same amount of food, warmth, protection and any other necessity. Thus, yielding their self-interests, cells began to differentiate.

As cellular differentiation evolved, bigger, stronger, and more diverse plants appeared. By allowing some cells to focus solely on the suction of water from the ground, and others to focus on photosynthesis, plants began to ascribe certain sections in the colony, not just certain cells, to dedicated tasks. This resulted in the emergence of *organs* such as root, stem, stalk, and leaves, and allowed for higher level plants to evolve. As before, the determining factor in the success or failure of a new evolutionary stage was the "consent" of cells or organs within the host system to yield their self-interest in favor of the interest of the entire system, in this case, a plant.

Animate

For some two billion years, plants were the rulers on planet Earth. But the desire to receive that broke Adam's *Partzuf* had more facets that needed correction, that is, to be taught how to work as a system, yielding selfish interest before the interest of the host system. As desires continued to emerge, those that correlated to Stage Three of the four stages began to manifest, creating more complex life forms.

Because of their higher level of desire, explains Ashlag in his "Introduction to the Book of Zohar," each specimen that belonged to Stage Three had a heightened sense of self-determination and a greater desire for autonomy. Thus, while specimens continued to recognize themselves as part of a species, they began to develop individual identities.[77]

Corals, for example, which evolved nearly 500 million years ago, were among the first species of animals to appear. Some of these developed (a primitive form of) muscles by which to stir their movement, and were thus able to move about relatively freely. Moreover, unlike plants, which provide for their nutritional needs using photosynthesis, corals must prey on other organisms to sustain themselves, and often contain algal cells to photosynthesize light for their supply of carbohydrates (sugars) (Image no. 3).

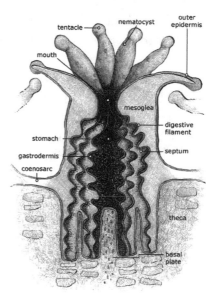

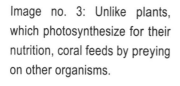

Image no. 3: Unlike plants, which photosynthesize for their nutrition, coral feeds by preying on other organisms.

But corals possess another form of tissue characteristic of animals: nerves. The appearance of a nervous system, particularly a Central Nervous System (CNS), allowed for enhanced control over the organism's function and facilitated the evolution of the diverse fauna that exists today.

Below is a very approximate timeline of the 3.8 billion year history of life on Earth, demonstrating how desires have manifested through evolution:

Still—Stage One
- 3.8 billion years since the appearance of simple cells (prokaryotes);

Vegetative—Stage Two
- 3 billion years since the appearance of photosynthesis;
- 2 billion years since the appearance of complex cells (eukaryotes);
- 1 billion years since the appearance of multicellular life;

Animate—Stage Three
- 600 million years since the appearance of simple animals;
- 570 million years since the appearance of insects;
- 550 million years since the appearance of complex animals
- 500 million years since the appearance of fish;
- 475 million years since the appearance of land plants
- 400 million years since the appearance of seeds;
- 300 million years since the appearance of reptiles;
- 200 million years since the appearance of mammals;
- 150 million years since the appearance of birds;
- 130 million years since the appearance of flowers;
- 65 million years since the non-avian dinosaurs died out;

Human (Speaking)—Stage Four
- 2.5 million years since the appearance of the genus Homo;
- 200,000 years since the appearance of Homo sapiens.

As we can see in the list above, evolution of the species and evolution of desires correspond rather nicely. The next chapter will be dedicated to the appearance and evolution of Stage Four in the desire to receive on earth—"the speaking"—which is the human being.

5

GENUS HOMO

Just as Stage Four is a natural evolution of the desire to receive, its corporeal parallel, human beings, appeared through a natural process of evolution following the same principles explained in the previous chapters. The genus Homo (humanoid ape) first appeared approximately 2.5 million years ago, and evolved as all other species do, by natural selection. As with animals, hominids that were healthier and stronger survived, and those that were less so perished.

Yet, hominids, and primarily the latest evolution of the species, Homo sapiens, invested far more energy and time on social relations than any other species. Albeit many species, such as dolphins, chimpanzees, and wolves, cultivate intricate social relations, social structures in human societies are dynamic and *evolutionary* by nature.

In that regard, Baal HaSulam wrote in the "Introduction to the Book of Zohar" that unlike animals, humans have the ability to sympathize with another's pains and joys, and animals do not.[78] In declaring this, Baal HaSulam was not referring to empathy as is often exhibited by animals between mother and offspring, and even among unrelated specimens of a species. Instead, here he speaks of an entirely new mechanism of the desire to receive: evolution through envy.

In item 38 of the introduction just mentioned, Ashlag explains the difference between desires in humans and in animals, and how envy increases our desires: "The will to receive in the animate, which lacks the sensation of others, can only generate needs and desires to the extent that they are imprinted in that creature alone."[79]

In other words, if an animal knows that eating is good, it may want to help another animal obtain food, as well. "But man," continues Ashlag, "who can feel others, becomes needy of everything that others have, as well, and is thus filled with *envy* to acquire everything that others have."[80]

Hence, even when we have had our fill of food, shelter, and all other essentials, our envy constantly compels us to want more: a greater house, stronger/healthier/more beautiful children (and preferably all the above), a bigger lot of land... the list is as long as the list of human desires. In that regard, Ashlag quotes the 1,500-year-old text of the Midrash, "One who has one hundred, wishes for two hundred, so the needs forever multiply until one wishes to devour all that there is in the entire world."[81]

Indeed, the appearance of Homo sapiens marked what appears to be a shift in the direction of evolution. Homo sapiens, it seems, were focusing not on developing a stronger, more adapt and agile physique, but on developing their intellect, and even more surprising, self-expression. Today, as Twenge and Campbell have shown in the above-mentioned *The Narcissism Epidemic*, this has become an epidemic of self entitlement. Thus, we see how Homo sapiens is the earthly representation of Stage Four in the desire to receive—the desire to become omnipotent and omniscient.

THE BEGINNING OF EGO

Ashlag's words quoted above mark a turning point not just in the history of human evolution, but in the evolution of the universe, as well. The (uniquely human) evolution-by-envy has shifted the very direction of evolution. Until the emergence of human ego, creatures evolved successfully if their internal organs cooperated, following the principle of relinquishing self-interest in favor of the system's interest, and leaving the system to care for their well-being.

Yet, it is important to note that the rule of Relinquishment of Self-interest in favor of the interest of the system applies not only to organs and tissues within a creature. Organisms do not exist in vacuum; they are branches, as we said in the previous chapter, of roots that appeared in the spiritual realm. For this reason, they operate in the same way that spiritual systems operate— yielding self-interest before the interest of the host system—or succinctly: altruistically. Their host system—the ecosystems in which organisms live—abide by the same rule, since no other rule enables life to perpetuate.

For this reason, the rule of yielding self interest we have been mentioning throughout the book applies just as rigorously to the creature's functionality within its environment. Thus, if a creature's physique works fine under certain environmental conditions, but conditions change, this creature's physique might become inadequate and even inferior to that of creatures with a less sustainable internal structure, yet higher adaptability to their environments.

Apparently, such was the case with the extinction of dinosaurs. For 165 million years, dinosaurs ruled the earth. But approximately 65 million years ago, they disappeared within a relatively short time. Theories as to the reason for their disappearance abound, but no conclusive answer has been found.

One possibility is the meteorite theory. According to the U.S. Geological Survey (USGS), "There is now widespread evidence that a meteorite impact was at least the partial cause for this extinction."[82] But while there is no scientific consensus around a meteorite impact being the cause, there is indeed consensus that, as published by the University of California Museum of Paleontology, "There was *global climatic change*; the environment changed from a warm, mild one in the Mesozoic era [era of dinosaurs] to a cooler, more varied one in the Cenozoic era [era of mammals]."[83]

Thus, whether it was a meteorite or something else that changed the climate, there was an abrupt change of environment to which dinosaurs (and approximately fifty percent of the species living on earth at the time) could not adapt. And so, they became extinct.

To survive, dinosaurs and almost all other animals must abide by the same law regarding their environment as their internal organs do: yielding self-interest in favor of the system's interest, in return for the system's care for them. When the rule is breached in the entire ecosystem, even if not willfully on the part of the animals, extinction occurs on a colossal scale simply because they did not adapt quickly enough.

A more recent, and far more successful example of animal adaptation to changing circumstances was reported by Swanne Gordon of the University of California in an essay titled, "Evolution Can Occur in Less Than Ten Years," published on *Science Daily*. "Gordon and her colleagues studied guppies—small fresh-water fish (Image no. 4) biologists have studied for long. They introduced the guppies into the nearby Damier River, in a section above a barrier waterfall that excluded all predators. The guppies and their descendents also colonized the lower portion of the stream, below the barrier waterfall, that contained natural predators. Eight years later..., the researchers found that the guppies in the low-predation environment... had adapted to their new environment by producing larger and fewer offspring with each reproductive cycle. No such adaptation was seen in the guppies that colonized the high-predation environment... 'High-predation females invest more resources into current reproduction because a high rate of mortality, driven by predators, means these females may not get another chance to reproduce,' explained Gordon. 'Low-predation females, on the other hand, produce larger embryos because the larger babies are more competitive in the resource-limited environments typical of low-predation sites. Moreover, low-predation females produce fewer embryos not only because they have larger embryos but also because they invest fewer resources in current reproduction.'"[84]

Image no. 4: Trinidad Guppy, the type used for the Damier River expermiment (Image: Photobank Lori)

In some cases, when required in order to increase their chances of survival, organisms (albeit only a virus in this case) may even "devolve" themselves. Such was the case with the Myxoma virus and the European rabbits in Australia (Image no. 5). Some 150 years ago, two dozen rabbits were released to the wild in Australia in the hope that they would reproduce enough to be hunted for sport. But the rabbits reproduced so successfully that within a few decades, they threatened to disrupt the wildlife equilibrium in the entire Australian continent. Wendy Zukerman, a reporter for *New Scientist Magazine*, published a detailed description of the episode on ABC Science. In her report, she writes, "By the 1920s, Australia's rabbit population had swelled to 10 billion."[85]

Image no. 5: European Rabbit in Australia
(Image: Photobank Lori)

Australian authorities made vigorous attempts to quell the rabbit population, but not until 1950 were they successful. In that year, continues Zukerman, "The biological control agent, Myxoma virus, was introduced to Australia's mainland."[86] As a result, "Myxomatosis [the disease caused by the virus] caused enormous reductions in rabbit numbers. In some areas 99 per cent of the rabbits were killed."[87]

But instead of extinguishing the European rabbits in Australia, their population gradually stabilized and even rebounded in some areas. Clearly, the virus had become less effective. When researchers looked into the reason for the virus' diminished impact, they discovered that it had mutated into a *milder* form that killed only 40 per cent of rabbits infected. Thus, researchers concluded, because the virus' only hosts were rabbits, it mutated into a less aggressive type, which guaranteed the survival of the rabbits, and as a result, the persistence of the virus, as well.

By weakening itself, the virus seemingly acted against its own interest, giving the rabbits' immune systems a better chance at fighting it. But the actual result of its self-induced weakening was the assurance that it would have a host in which to dwell for generations to come. Indeed, to this day, myxomatosis is responsible for many deaths among rabbits, but not enough to altogether extinguish them. It seems as though rabbit and virus have achieved equilibrium, and hence co-existence.

MAN—THE ONLY EXCEPTION

In the previous section, we saw how the rule of Yielding Self-interest in favor of the system's interest in return for the system's care, applies not only to all organisms, but also to the organism's functionality within its habitat (ecosystem). Yet, there is one exception to the rule: man. To understand why man is different from all other animals, we need to reflect on the four stages. Stages One through Three reflect desires to receive pleasure from a giver, either by receiving pleasure directly from it or by returning its pleasure. But Stage Four is essentially different: it reflects a desire Put differently, Stage Four wishes to attain a goal that is, by definition, unattainable. Just as a son cannot be his father, Stage Four cannot be Stage Zero. But just as a son can be *like* his father, Stage Four can be like Stage Zero.

Being a desire to receive, and knowing that being like Stage Zero, the Root, is the highest possible reward, this is what Stage Four wishes to achieve. As a result, we—its corporeal personification—strive to achieve the same. Subconsciously, our desires for fame, power, wealth, erudition, and immortality are really desires to become godlike. No person escapes these desires, since we are all parts of Stage Four, which was broken along with Adam's soul. The only variations among humans are in the intensity and proportion of these desires, but not in their components.

Evidently, there are people whose desires for fame, fortune, and brilliance are very small—these are simple folk content with shelter, family, and very basic sustenance. In such people, the desires of Stage Four are less dominant; hence, they will have less ambitious goals. But even in the most sedate individual there is a "devil" that wishes for a little more than one's neighbor possesses. These are the desires of Stage Four—the sense of entitlement that Twenge and Campbell write about—and they are almost uniquely human.

These desires are also what make us the exception to the rule that governed evolution until the emergence of Homo sapiens. Because humans possess an innate aspiration to become like the Creator, we tend to be active in our approach to challenges, rather than passively adapting to conditions, as do other animals. Hence, instead of adapting our bodies as best we can to changing climates or to threats, we try to change the climate or to eliminate the threats.

One such effort was changing our "personal microclimate," our immediate surroundings, by covering our skins with those of animals, whose fur provided better protection against the elements than our own. And instead of relying on our (clearly insufficient) physical strength to catch our food, we developed increasingly sophisticated tools to assist us in hunting, as well as for protecting ourselves against prey animals. Today there is unequivocal evidence that primates, some mammals, and even birds use tools such as rocks, twigs, and branches to assist them in acquiring food and in fighting. But systematic tool and weapon production, such as carving stones and bones into spears, is a uniquely human ability (Image no. 6).

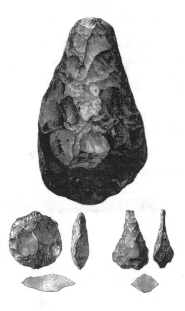

Image no. 6: Hand axes from Kent (England), made during the Lower Paleolithic period (Old Stone Age), 2.5 million – 200 thousand years ago.

Another very important discovery that early humans (Homo erectus) made was the control of fire. Fire allowed humans to keep their habitat warm, deter prey animals, and even cook. The discovery of ways to make and to control fire marks a dramatic shift in evolution. Man was now an animal that could change its environment to fit its needs, instead of having to change itself to fit the environment.

According to a document titled, "The Great Ice Age," released by the U.S. Geological Survey, "The Great Ice Age ... began about a million or more years ago."[88] The vast ice sheets allowed humans to migrate from Africa and to gradually spread all over the world. With fire and clothing, they could sustain themselves in less hospitable climates and thus became the most adaptable and ubiquitous mammal on earth.

BODY VS. MIND

A deeper and far more important aspect of the shift in evolution that the appearance of man represents is that unlike other animals that develop their bodies, humans develop their *minds*. To cope with danger or to obtain food, animals try to outrun or outfight their attackers or prey.

Humans, instead, build weapons. To cope against the cold, animals grow thick fur and layers of hypodermal fat. Humans light fires.

The use of the intellect instead of the body to obtain desirables also allows humans to *plan ahead*. While some animals store food for the winter, only humans *grow* food and clear unwanted vegetation from the land to make room for plants that serve *them* as food. According to most researchers, agriculture began between 10,000 and 15,000 years ago in the Fertile Crescent (although new data collected by a team led by Dr Robin Allaby from the University of Warwick has found evidence that plant agriculture began in Syria as early as 23,000 years ago).[89]

Although man's ability to grow food may seem much ado about nothing today, when humans first began to cultivate land, they, in a sense, became creators—they began to change their environment. This is a feat that only a desire of Stage Four can conceive.

Yet, with progress comes problems. All creatures, except man, must adhere to the rules of their ecosystem or they will perish. Man is the only organism that can plan and execute change in its environment at will. When this happens, man must learn the rules by which ecosystems work, or the changes might prove to be disastrous to the ecosystem, and by consequence, to its inhabitants, man included.

In Chapter 4, we said that in the human body, as in any organism, each cell has a particular role. Also, we wrote, "For the organism to persist, each cell must perform its function... and yield the goal of maintaining its own life before the goal of maintaining the life of its host organism. If a cell begins to act contrary to that principle, its interests will soon clash with those of the body and the body's defense mechanisms will... destroy it."

Similarly, when man became potent enough to alter his ecosystem, he had to learn how to behave like a cell in an organism—refraining from jeopardizing the system's sustainability, and risk having the system need to rid itself of the danger by either eradicating the human race altogether or by dying itself, killing the human race in the process, as described in regard to cancer. Today, I believe it is quite evident that Nature is already "taking compensatory measures" to balance humans' detrimental actions.

But ten or so millennia ago, things were very different than they are now. Homo sapiens were just beginning to enjoy the benefits of knowledge and technology and the concept of humans risking their habitat was not on anyone's mind. The development of agriculture shifted people's lifestyles from hunting and gathering to a more sedentary comport, one of which consequences of which was the acceleration of technological development.

Another important issue that *was* on people's minds at that time (and still is for many) was religion. Prof. Jared Diamond, acclaimed author of *Guns, Germs, and Steel: The Fates of Human Societies*,[90] said in a lecture titled, "The Evolution of Religions" at University of Southern California,[91] that approximately ten and a half thousand years ago, religion changed its functions. He explained that religion had adopted a role of explaining things. Religion began to explain all that was unknown and unfamiliar, and thus provided solace and confidence to people.

But the important thing to note about religion at that point is not so much the direction in which it developed, but the very *fact* that it developed. The existence of an institutionalized, organized entity that provided answers meant that people were beginning to ask questions—profound questions about the purpose of life and the laws that govern it. This later prompted the emergence of Kabbalah, precisely in that same area—the Fertile Crescent —as we saw in Chapter 1.

In addition to the evolution of religion, because the agricultural advances we just mentioned encouraged people to abandon their nomadic lifestyles for a more sedentary one, the population in the Fertile Crescent began to grow. And when technological developments, such as the invention of the wheel, encouraged further development and urbanization, more organized forms of government and religion ensued. Thus, Mesopotamia gradually became what we now call "The Cradle of Civilization."

6

In Opposite Directions

As mentioned in Chapter 1, Mesopotamia, The Cradle of Civilization, was also the birthplace of Abraham, the harbinger of Kabbalah. The conflict between Abraham and Nimrod, ruler of Babylon, stands for much more than a conflict between a ruler and a defiant subject. It is a conflict of *perceptions*. To Nimrod, reality is a "federation" of forces that he must please, serve, and appease by sacrifice. To Abraham, there is only one force, and worshiping it means living by its law—the law of giving, as simple and as straightforward as that. Considering this contrast of views, it is no wonder that Nimrod had to either destroy Abraham or expel him.

But Abraham's departure from Babylon did not quiet the polis. The trends that had prompted Abraham's search for life's secret continued to intensify and to spread through the bustling city, fueled by the same forces that power the process of evolution. Yet, in Babylon, these trends began to manifest a conduct that is uniquely human—egoism.

Baal HaSulam explains that egoism is a natural trait for humans. He declares that it is human nature, and that Kabbalah offers a way to turn its evident detrimental consequences into positive ones. In "Peace in the World," he writes, "In simple words we shall say, that the nature of each and every person is to exploit the lives of all other people in the world for his own benefit. And all that he gives to another is only out of necessity; and even then there is exploitation of others in it, but it is done cunningly, so that his neighbor will not notice it and concede willingly." [92]

But before we delve into the solution that Kabbalah offers to human egoism, we need to understand how the desire to receive, initially created by the desire to give—the Creator—has become egoism. "The reason for it," continues Ashlag, "is that... man's soul [desire] extends from the Creator, who is one and unique. ...Hence, man, too... feels that all the people in the world should be under his governance," [93] just as the whole of nature is governed by the law of bestowal, the Creator.

Moreover, unlike all other elements in Nature, which are forced to behave in congruence with their environment, human beings have the power to change the environment. This gives us something that no other creature has: free choice. Put differently, human beings can choose to be like the Creator—giving—and acquire the power and cognizance that come with it, or remain as we were born—self-centered and limited.

When the stages of desires cascaded from the desire to give, the desire to receive evolved with each new stage. In the physical world, too, the evolving desires manifest in the different stages of evolution (Figure no. 9): At the bottom of the pyramid are the minerals and the inanimate materials. This is the Still Level, corresponding to Stage One. Above that is the flora—corresponding to Stage Two, topped by fauna—Stage Three, and above all is man (speaking)—Stage Four.

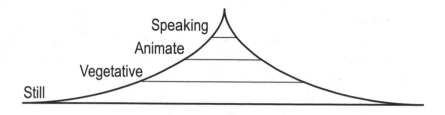

Figure no. 9: Pyramid of Desires. The top of the pyramid is also the part that governs it, and hence the part that has free choice in how to do it, and the responsibility to do it right.

Considering that all that exists are the desire to give and its offshoot, the desire to receive, it is evident that the speaking level (us), possessing the most intense, sophisticated and complex desire to receive, is not just an inseparable part of Creation, but is its apex and governor. And just as the brain governs the entire body, yet is also completely dependent on it for its survival, we must learn how to govern and nurture the whole of the pyramid of Creation if we are to survive.

A PYRAMID WITHIN A PYRAMID

The reason why Abraham was the only one of his generation to discover life's creative force is that he was a piece of Adam's *Partzuf* that was ready to reveal it. But the goal of Creation is not for only one person to achieve the Creator-like state, but for *all* of humanity to achieve it. Hence, Abraham's discovery was not a one-time-thing, but an antecedent to a new stage in the spiritual evolution of humanity.

Abraham realized that life is a pyramid whose peak is the Creator's trait of bestowal. He also realized that human desires would only intensify, as they have done since the dawn of Creation. And finally, Abraham knew that this awareness, along with having the correction method provided by Kabbalah, were the only ways to avert the collapse of the system due to the heightening egoism. But in the absence of tangible proof, only a handful followed Abraham and united around the goal of attaining the Creator. When those who went with him grew and became a nation, they were named after their goal: *Ysrael* (Israel), from the Hebrew words *Yashar El* (Straight to God).

Historically, Babel did not collapse immediately or even soon after Abraham's departure. It continued to fluctuate in dominance and prominence for more than a millennium following his leave, including the resettlement of Hebrews in Babel after their exile following the ruin of the First Temple. However, from the spiritual, Kabbalistic perspective, Nimrod's triumph in Babel sealed its doom because it perpetuated the rule of egoism rather than altruism.

THE CURE THAT WASN'T

In truth, Abraham's method was very simple: in the face of heightened egoism, unite and thus discover the quality of bestowal—the Creator. As we have been showing throughout the book, every element in nature behaves in this way. The initial levels of desire to receive require very limited organization and form small systems where each element dedicates itself to its host system. We call these elementary systems, "atoms." The more evolved levels of desires place atoms within systems we call "molecules." As the desire evolves further, these systems organize within even bigger systems called "cells." These group into multicellular creatures, finally leading to the creation of plants, animals, and humans.

In all of this, there is only one principle: the desire to receive in all the elements wishes to receive, and the only way to create balance and sustainability in the system is to unite under a higher-level system. This is what Abraham's method sought to *consciously* emulate.

As we have said, the desire to receive in humans becomes egoism because of our sense of uniqueness. Hence, the antidote to egoism is the exact same cure applied by Nature—the construction of a system to which all parts will contribute and yield their self-interests. In return, the system will guarantee the well-being and sustainability of its elements. Scientists today wish to discover the conditions that existed in the early universe by recreating those conditions on a miniature scale in facilities such as the CERN Hadron Collider in Switzerland. Similarly, by imitating Nature's "natural" conduct, we will discover its law of bestowal.

In truth, the *modus operandi* is really quite simple: If you think like a giver and act like a giver, we have to at least consider the possibility that you have a small amount of giving in your nature, to paraphrase Douglas Adam's celebrated quote from *Dirk Gently's Holistic Detective Agency*.[94]

However, Nature does not provide us with the instincts to emulate it, as it does with the rest of its elements. Because we are meant to be its rulers, our task is to *study* these rules by ourselves and subsequently implement them. This is why, when Nimrod expelled Abraham, the only man who could teach this rule to the Babylonians, he also denied his people the method of achieving unity—the one antidote to the growing egoism and alienation among his people.

Following Abraham's departure, Babel continued extolling self-centered abandon. But although pleasure and enjoyment do not contradict the purpose of Creation—as we know from Stages Three and One, which received the Creator's pleasure—receiving pleasure is neither the ultimate goal nor the greatest delight. Man's greatest delight and ultimate goal are to become like the Creator, and the Babylonians' negation of that goal is what ultimately ruined them. While Israel was forming into a nation, as described in Chapter 1, Babel experienced violent vacillations as the unbridled egoism of its people intensified. Its final disintegration in the 4th century B.C. proved a long, yet unavoidable process.

Yet, Babel was only the first stage in building the highest level in the pyramid of desires—the speaking level. As with all other elements in Creation, the final level in the pyramid consists of a root and four stages of evolving desires. Abraham is considered the Root Stage, hence his epithet, *Avraham Avinu* (Abraham Our Patriarch), referring to him being the progenitor of the nation that strived to reach the Creator. Later, as we know, he became known as the father of all three Abrahamic faiths, the monotheistic religions of Judaism, Christianity, and Islam.

As desires kept evolving in humanity, a new level of desire in the pyramid emerged atop the Root level, approximately when Egypt was in its prime. This level corresponded to Stage One, and as the Root Stage had its harbinger, Abraham, Stage One had a harbinger of its own, Moses. And just as Abraham was forced by Nimrod to exit Babel, Moses had to flee Pharaoh and exit Egypt, as described in the Pentateuch, "But Moses fled from the face of Pharaoh, and dwelt in the land of Midian" (Exodus 2:15). To understand the importance of Moses' mission, we need to understand a concept that initially appears to be unrelated— the concept of free choice, as explained by Kabbalah.

FREE CHOICE

Asalready discussed, the evolution of humanity corresponds to Stage Four in the evolution of desires. In this stage, the desire to receive realizes that behind all that occurs is a thought, a purpose that dictates this series of changes. In our lives, this translates into a child's drive to not only emulate its parents' actions, but to wish to *know* what they know.

To obtain the Creator's thought, Stage Four needs freedom of thought and freedom of will so it can develop its perceptions independently. Similarly, if you teach a child to think and view the world through a narrow perspective, he will make a very loyal soldier, but probably not a great strategist or general. This, also, is the reason why children—especially in early childhood, before we accustom them to indolence—wish to do things by themselves instead of letting their parents do it for them.

Thus, the need for free choice requires our ignorance of the law by which all creatures achieve balance and sustainability through yielding self-interest to the interest of the host system, so that we can discover it for ourselves. If we knew that this was the law in effect, and that it is as rigid as the law of gravity, we would not dare defy it. And if we had no choice but to follow it, we would, at best, become obedient children, but we would remain children, forever inferior to the desire to give that created that law.

To equal the Creator, we must learn how to "build" Creation by ourselves, every element within it, the reason for its existence, how and when it emerged, and if and when it will expire. To learn that, evolution has created the perfect infrastructure for our learning: it has built a universe in which every element abides by the law of yielding self-interest in favor of the system's interest. Additionally, evolution denied us the knowledge of this law, and gave us the power to act contrary to it, or not, depending on our choice. And most of all, evolution did not reveal to us the reward for observing this law.

Cells in the body sympathize with the life of their host organism, not their own. If this were not so, they wouldn't be able to operate in its favor and would become malignant or even prevent the initiation of life altogether. This sympathy is so complete that cells are even willing to terminate their own lives to promote the growth of the entire body in a process known as "apoptosis" or "Programmed Cell Death" (PCD). In embryos, for example, the embryo's shape of feet is determined by apoptosis, which finalizes the differentiation of fingers and toes when cells between the fingers are deliberately put to death by their hIn return for the cells' sympathy, they are "rewarded" with the perception of the world of their host organism, instead of their own. That is, cells behave as though they are equipped with an innate perception of the entire organism of which they are parts. If they did not function in this way, they would instinctively try to fight their neighboring cells for supply of nutrients and oxygen, as do unicellular creatures. When such a malfunction occurs in a cell within an organism, this can develop into cancer.

If we, like cells in an organism, could sympathize with our host system—Planet Earth—but even more than that, with the forces that built and sustain the Earth, we would obtain the broadest possible perception and transcend such concepts as time, space, and life and death as we know them. Our perception would reveal that we are part of a much broader system than our immediate surroundings, just as cells are part of the entire organism. In that state, we would be able to think and act as the Creator—the desire to give. And in achieving this, we would achieve the purpose of Creation—becoming Creator-like.

Yet, if we could see that by yielding our self-interest we are rewarded with being Creator-like, we would do it in order to receive pleasure, without the aim to give, and without the aim to give we would remain self-centered, disparate from the Creator. To achieve a Creator-like state, we must choose it *freely*, without being lured in any way toward altruism. Because, as we explained about the four stages, the aim to give is what makes us Creator-like, the desire to receive must not feel that we will receive pleasure or benefit in giving, so it would not create selfish motivation.

When we understand that, we will understand how important the restriction of pleasure by Stage Four is to us. If Stage Four did not repel it, we would succumb to the pleasure just as a baby enjoys its parents' strength and benevolence toward it, and we would not be able to become like the Creator. Instead, we would be taken by the pleasure, just as moths are lured by the light of a lamp on a dark night.

IN THE FACE OF EVOLVING DESIRES—UNITE

Earlier, we said that when desires evolve in Nature, they create increasingly complex structures. Each new level rises to a higher degree of desire to receive when creatures of the current level join to form an aggregate of collaborators. By so doing, the creatures of the current (and presently highest) level create a system to which they can yield their self-interests, which provides them with sustainability and adherence to Nature's law of giving. When this happens in humans, we, too, start from the smallest structure—a single person—and work our way toward increasingly complex societies. The only difference is that we must create these social structures that adhere to the law of giving by ourselves.

Abraham's family was actually the first group to create that system, and then harness its members into a system whose parts were united by dedication to their host system. As Maimonides narrates (Chapter 1), this initial system grew into a group. Yet, only in Egypt—when their number sufficed—did the system grow into a nation. When Moses brought Israel out of Egypt, the family of seventy that had gone into Egypt now consisted of several millions (there are many views on precisely how many came out of Egypt, but the common figures are between 2 and 6 million men, women, and children, excluding the mixed multitude).

Clearly, Moses' job was far more challenging than Abraham's. He could not gather the entire nation in his tent, as did Abraham with his family and few disciples, and teach them the laws of life. Instead, he gave them what we refer to as the *Five Books of Moses*, known in Hebrew as the Torah, which means both "Law" (of bestowal) and "Light." In his books, Moses provided depictions of all the states that one experiences on the way to becoming like the Creator.

The first part of the way to emulating the Creator was to exit Egypt, venture into the Sinai Wilderness, and stand at the foot of Mount Sinai. According to ancient sources, the name, "Sinài," comes from the Hebrew word, *Sinaa* (hatred).[95] In other words, Moses gathered the people at the foot of Mount Sinai—the mountain of hatred.

To interpret the mountain-of-hatred allegory, Moses' teachings showed the people how hateful they were towards each other, how remote they were from the law of bestowal. To reconnect with the law of bestowal, or the Creator, they united, as described by 11th century commentator and Kabbalist, Rashi, "As one man in one heart."[96]

Baal HaSulam elaborates on this process in his essay, "The Mutual Guarantee,"[97] where he explains that in return for their pledge to care for each other, Moses' people were given the Torah. They attained the law of bestowal and obtained the light, the altruistic nature of the Creator. In his Baal HaSulam's words, "Once the whole nation unanimously agreed and said, 'We shall do and we shall hear,' ...only then did they become worthy of receiving the Torah, and not before."[98]

Now we can see how important Moses' mission was, and why free choice is a prerequisite to accomplishing it. The leaders of Abraham's group were all family and were naturally united. But Moses had to unite a *nation*. To achieve that, the entire nation had to agree on a path. By making a free choice to unite, despite the evident egoism (allegorically described as "standing at the foot of Mount Sinai"), a nation was admitted into the law of giving. This was the first time in the history of humanity that people en masse attained the quality of the Creator, and from this point forward, choosing unity in the face of growing egoism will be the only way to achieve the Creator.

THE OTHER WAY

The sages of the Talmud wrote, "One who has one hundred, wishes for two hundred."[99] Since the dawn of Kabbalah, its practitioners stated that our desires evolve. They grow in both intensity and quality, meaning not just how much we want, but also *what* we want. Eventually, these desires evolve to become the ultimate desire—to be like the Creator.

But Kabbalists have also stated that we have free choice in how we arrive at the greatest desire, which also yields the greatest pleasure. They said that there are two ways to reach this goal.

1. We follow Moses' example and unite. We do that by studying how Nature works at its most fundamental levels, how we, being offshoots of the law of Nature operate, and then try to work like Nature, in unity, just as a child imitates its parents.

2. We ignore the available information and try to discover the secret to a good and sustainable life by ourselves. This can be compared to a child sitting behind the wheel of a car but is too small to see out the window. Naturally, this will result in recurring accidents with horrific consequences.

Kabbalists called the first, enlightened way, "The Path of Light," and the second, torturous way, "The Path of Suffering."[100]

The evolution of desires occurs irrespective of our choices. When it is not accompanied by a calculated effort to unite and to choose the path of light in order to discover the law of giving, there is nothing to regulate the growing desire and funnel it in constructive directions. The result is increased and unchecked egoism. This is usually accompanied by "an accident"—disintegration and defeat as it happened in Babylon and in Egypt.

Indeed, the history of the Israeli nation is the best example of this statement. As long as they followed Abraham's teaching, they succeeded. When they did not, they were defeated and exiled.

Approximately 1,900 years ago, a new level of desire to receive emerged. This required a renewed effort and a renewed choice to unite. Yet, the people of Israel were not ready to make the effort. Instead, they fell into hatred and egoism. The Babylonian Talmud, written around the 5th century C.E., explains that the sole reason for the defeat of Israel and the ruin of the Temple was unfounded hatred.[101]

Since that ruin, the world has had only one path to tread— the path of suffering. The path of light was known to very few individuals in the generations that followed, and every few decades they would warily try to expose it. But seeing that people were not yet ready to contemplate the truths that that path held about reality, they kept it to themselves and to those rare devoted students who sought the truth at all cost.

Yet, as we will see in the next chapter, the years of obliviousness to Kabbalah were not in vain. They have given us much knowledge and myriad observations of Nature as a whole, and of human nature in particular. Without those years, the resumption of acceptance of the knowledge that Kabbalah provides would not be possible.

7

THE GREAT MINGLING

The first centuries in the Common Era were a tumultuous period in the history of Europe and the Near and Middle East. The Romans conquered large parts of Europe, North Africa, and the Near East (including what is now considered the Middle East). Additionally, Judea was conquered (by Rome), then rebelled, lost, and the Jews were exiled. Christianity, too, was making its debut, and Britain was conquered by Emperor Tiberius Claudius. As we will see in this chapter, the exile of the Jews and their dispersion throughout Europe are tightly connected to the evolution of desires.

During those first centuries, a new and different world was forming. Having been exiled, Jews were spreading throughout the Near East and Europe, and Christianity was gradually taking hold, becaming the official religion of the Roman Empire when Emperor Constantine the Great adopted it in the 4th century.

When Islam was promulgated in the 7th century, it created a situation where the majority of the people in Europe and the Near and Middle East adhered to one of the three Abrahamic faiths. Today, this may not seem extraordinary. But in those years, this shift of faith was a revolution caused by the emergence of the next stage in the evolution of desires—Stage Two.

Stage Two, the emergence of the desire to give within the desire to receive, instigated a crossing of two paths—that of Israel with that of all other nations. For the first time since Abraham departed Babel and formed the group that aimed *Yashar El* (straight to God), which evolved into the nation of Israel, his message—love thy neighbor as thyself—was being heard en masse. Because Stage Two—the desire to give—was beginning to manifest, the message of giving and compassion could now be heard, though it was clearly not executed as well as it was taught.

This chapter will examine the sub-surface processes that unfolded between the writing of *The Book of Zohar* (also called *The Zohar* for short) in the 2nd century C.E. and the writing of *The Tree of Life* in the 16th century. These dates (very) roughly parallel the period between the Roman conquest of Judea and the onset of the Renaissance, or what we now call "the Middle Ages." As with the rest of the book, the goal is not to focus on particular events, but to provide a "bird's-eye" view of history, showing how processes correspond to the evolution of desires. In the case of the time frame just mentioned, it is probably best to begin with the Roman conquest and the ruin of the Second Temple.

THE DISPERSION OF JUDEA

The defeat of the Jewish revolt against the Romans (66-73 CE) caused the ruin of the Second Temple and the dispersion of Judea. (The First Temple was built by King Solomon in the 10th century BCE, and was ruined by the Babylonians in 586 BCE.) This dispersion signified something far more important than the conquest of one nation by another. It reflected the extent of the Israeli nation's spiritual decline. The Hebrew word *Yehudi* (Jew) derives from the word *Yechudi* ("united," or "unique"), referring to the state of the Israeli nation of the time: perceiving (and adhering to) the unique force of bestowal that governs all of life.

Yet, as we explained in previous chapters, the desire to receive is an ever-evolving force and requires constant adaptation. Constant effort is required to harness the newly emerging desires to work in unison—with the intention to bestow, and adhering to the law of yielding self-interest in favor of the interest of the host system. And because the desires evolve, the means to harness them must evolve accordingly.

As explained in the previous chapter, unlike animals, humans must constantly realize their place in Nature and *choose* to be constructive parts of it. However, if we act to the contrary, the negative outcome will not be immediately evident. This leaves us room to maneuver and to calculate.

At the same time, if we choose to act in accord with Nature's law, we will not immediately notice the positive result. Thus, because the reward and punishment are not immediately discernible, if we choose to do so nonetheless, it will be only because we want to discover Nature's law of unity and giving, and not because we expect an immediate reward. In this way, we act out of an *intention to become givers* instead of out of our inherent *desire to receive*.

But during the first century CE, the evolution of the desire to receive prompted the emergence a new level of desire. Until the arrival of that level, the Jews that returned from the exile in Babylon—after the ruin of the First Temple—kept their unity and their perception of the cohesive law of life.

In truth, only two of the twelve tribes returned from their Babylonian exile because the level of egoism was also growing among Israel, and the majority of the tribes could not resist the egoistic drives within them. These drives separated them from the nation of Israel, which consists, as explained, of people who live by the law of unity, and not of genetically related individuals. But when Stage Two in the evolution of desires began to manifest in Israel, even those who returned from Babylon could not maintain their altruism. Instead, they fell prey to their self-centered desires.

The Babylonian Talmud explains that the sole reason for the defeat of Israel and the ruin of the Second Temple was unfounded hatred: "The Second Temple, why was it ruined, since they engaged Torah and *Mitzvot* [spiritual learning] and in good deeds? It was because there was unfounded hatred in it."[102] In the absence of unity, and because many Jews wished to emulate or even join the Roman culture, the Jewish revolt was hopeless from the start.

Still, even after the revolt, many among Israel maintained their cohesive perception of reality. Rabbi Akiva, for example, whose Talmudic epithet was "Head of all the Sages," lived and taught in the years following the ruin. According to the Babylonian Talmud, Rabbi Akiva had 24,000 students, but they, too, died (according to the Talmud) because they were not united.[103]

Of the 24,000 students, only four survived. And of those four, two became the greatest sages of their generation, and possibly of all time. The first was Rabbi Yehuda, known as Rabbi Yehuda HaNasi (the president), who became president of the Sanhedrin and chief redactor and editor of the Mishnah, the corpus that is the foundation on which both parts of the Talmud are built. The other student was Rabbi Shimon Bar-Yochai (Rashbi), who became known as the author of *The Book of Zohar* [*The Book of Radiance*]—the seminal book of Kabbalah, which all Kabbalists study to this day and from which they all derive their wisdom.

Through the centuries, there have always been sages who kept the wisdom vibrant and evolving. They understood the nature of the desire to receive and produced texts that interpreted *The Zohar*, as well as other books of Kabbalah. Yet, for the most part, their books—written from the Kabbalistic-altruistic perception of reality—were misunderstood by all except for fellow Kabbalists because they were read from an egoistic perception. This prevented readers from grasping the true meaning of the texts. In much the same way, a person who is blind from birth cannot understand the meaning of vision, much less the joy that comes from observing a beautiful landscape or the captivating power of the view of an ocean's stormy shore.

Thus, because of the decline of the spiritual perception (altruism) among Israel, Abraham's dream of teaching the entire world the single law of existence had to be postponed until people were once again ready to learn about this law. *The Zohar* was concealed soon after its completion and remained hidden for more than a millennium. Kabbalists, too, cloaked the wisdom in mystery and misconception, and declared that only those who met rigorous conditions were permitted to study it. Since they knew that the majority of people were too far removed from spiritual perception to properly grasp the concepts of Kabbalah, Kabbalists distracted people's minds with stories of miracles and charms, and set up boundaries such as age, sex, and marital status to deter would-be students from probing Kabbalah.

Indeed, the misperceptions of Kabbalah were so deeply rooted that even after the reappearance of *The Zohar* (Image no. 7) in 13th century Spain in the possession of Rabbi Moshe de León, the book was often misunderstood and considered abstruse text until such Kabbalists as the Vilna Gaon (GRA), Rabbi Isaac Safrin, and others offered clearer interpretations. Even so, it was not until the 1940s, when Yehuda Ashlag (Baal HaSulam) offered his complete *Sulam* (Ladder) commentary on *The Book of*

Zohar—with four explanatory introductions—that this profound composition could be properly studied and comprehended.

Image no. 7: The title page of a 1558 edition of *The Book of Zohar*, Mantua, Italy. The text begins, "The Book of Zohar on the Torah, from the Godly sage, Rabbi Shimon Bar-Yochai..."

But in the early post-ruin-of-Second-Temple years, the world was treading a very different route. The Romans were the empire in the Mediterranean, Near East, and Europe, and their (essentially Greek) culture and philosophy reigned. The Hellenistic perception of the world did not agree with that of the rebels from the land of Israel. Moreover, the majority of Jews did not agree with the principles of their forefathers, and abandoned them in favor of the ego-centered Hellenistic Greek-Roman culture.

That said, several renowned scholars of the renaissance believed that the Greeks did adopt at least some of their concepts from Kabbalah. Johannes Reuchlin (1455-1522), for example, the great humanist and political counselor to the Chancellor, wrote the following in his *De Arte Cabbalistica* (*On the Art of Kabbalah*): "*Nevertheless his* [Pythagoras'] *preeminence was derived not from the Greeks*, but again from the Jews. As 'one who received,' he can quite justly be termed a Kabbalist. ...He himself was the first to convert the name Kabbalah, unknown to the Greeks, into the Greek name philosophy."[104]

A predecessor of Reuchlin, Giovanni Pico della Mirandola (1463-1494), an Italian scholar and Platonist philosopher, wrote in his *De Hominis Dignitate Oratio* (*Oration on the Dignity of Man*), "This true interpretation of the law, which was revealed to Moses in Godly tradition, is called 'Kabbalah.'"[105]

But the principle that the Greeks did *not* adopt was the most important one of all: the intention to revoke self-centeredness in favor of system-centeredness *in order to become like the Creator*. The latter part of that phrase, the *reason* for shifting one's focus, is the reason why the wisdom Kabbalah was devised to begin with. Had the Greeks adopted it, history would have unfolded very differently.

Yet, it was through no fault of the Greeks that they did not adopt it. They did not know about it, as there were no Kabbalist teachers among them, and hence none who could educate them properly. Moreover, having heightened egos themselves, the Jews, too, were adopting the Greek-Roman ways, and those who were not were the Romans' fiercest enemies in Judea. In consequence, there was no one to show the Romans that they were missing anything that could be of value to them. And so the Romans pursued the Hellenistic culture until Emperor Constantine the Great adopted Christianity in the 4th centuThe Jews' adoption of the Hellenistic culture was no coincidence.

The establishment of the First Temple had marked the highest spiritual point (perception of the law of giving) in the history of the Israeli nation. From then on, a gradual process of decline was underway. The evolution of desires was affecting the Jews just as it was affecting all other nations. As a result, many of the Jews could not maintain their spiritual, altruistic perception of a unified force, and turned to more self-centered cultures that suited their egoistic perception.

Thus, the Babylonian conquest and subsequent exile of the Hebrews at the time of the First Temple were only a manifestation of their spiritual state at the time. And because of the waning spiritual state of the Hebrews in Babylonian captivity, only two of the twelve tribes that went into exile, Judah and Benjamin, returned. The ten tribes that remained in exile became so thoroughly mingled with the locals that they had completely forgotten their tenets, and their traces have been lost to this day.

Yet, the evolution of desires did not stop there. Judah and Benjamin gradually declined, as well, and the complete dispersion of the Jews was only a matter of time. Indeed, the Jews' loss of spiritual perception was a long process that spanned centuries, but its course was set. When the Romans finally conquered Israel and destroyed the Second Temple, Israel was already a nation whose majority did not want to maintain its spiritual (Kabbalistic) mindset and preferred the Hellenistic concepts in its stead. In consequence, they, too, were exiled and dispersed. And while many Jews remained in the land of Israel even after the Roman conquest, and compiled some of the most significant texts in Judaism, the Jews as a people were already spreading throughout Rome and subsequently Europe.

In *The Wars of the Jews*, Chapter 1, translated by William Whiston, Josephus Flavius describes the expulsion of the Jews by the Romans: "And as he remembered that the twelfth legion had given way to the Jews, under Cestius their general, he expelled them out of all Syria, for they had lain formerly at Raphanea, and sent them away to a place called Meletine, near Euphrates, which is in the limits of Armenia and Cappadocia."[106]

In Chapter 3 of the same book, Flavius elaborates: "For as the Jewish nation is widely dispersed over all the habitable earth among its inhabitants, so it is very much intermingled with Syria by reason of its neighborhood, and had the greatest multitudes in Antioch by reason of the largeness of the city, wherein the kings, after Antiochus, had afforded them a habitation with the most undisturbed tranquility."[107]

Thus, gradually, the Jews migrated throughout Europe and much of today's Near East. As a result, the history of the Jews and the history of Europe have become tightly linked.

THE AGE OF CONCEALMENT

The Middle Ages is a very peculiar period in history. Views on when it began and when it ended seem to range from 2nd-5th century to 15th-18th century respectively, depending on the researcher's field of expertise. Some mark the fall of the Western Roman Empire as its beginning and the fall of the Eastern Roman Empire as its end. Others see the beginning of the Middle Ages as the time when Emperor Constantine the Great summoned the First Council of Nicaea, in 325 CE, and its end as the time when Martin Luther was excommunicated (1521) and the Protestant Church was established.

Kabbalah does not define any age as being in "the middle," but it does consider the period between the writing of *The Book of Zohar* and the writing of *The Tree of Life* as a distinct period in the evolution of humanity. In a sense, the term, "The Dark Ages," would be more suitable to describe this period in history, since this is roughly the period during which Kabbalists concealed their knowledge and made it a secret teaching, known to only a few.

Within this period, and in accord with the "bird's-eye" view of this chapter, we will relate more to the *processes* that occurred between the writing of these books than to specific events. This should make it easier to see how desires, which on the human level appear more as ambitions, steer the processes that form the history of humanity.

In Kabbalah, the period between the writing of *The Book of Zohar* and writing of *The Tree of Life* has a crucial role. Without it, the purpose of Creation would not be achieved. To reiterate in a word, the purpose of Creation is for every person to know the Creator and become like it. Abraham's group was the first to achieve that. Yet, Abraham's goal was not only for his group to achieve it, but for *every person* in the world. Moses helped Abraham's cause by expanding the attainment of the group into the attainment of an entire nation.

But while Moses' accomplishment is truly momentous, there is still a long way to go before the final purpose is achieved. For the whole of humanity to attain the Creator, the law of bestowal, they must all *want* this to happen. And for that, all the people must sense that a) the road of egoism is unsustainable, and b) that there is another way—to recognize a previously undiscovered law in Nature, and learn to implement it.

During Stage Two in the evolution of desires, this unfolds in a fascinating manner. Israel, on one hand, declines from its altruistic state and falls into egoism. The rest of the nations, on the other hand, discover the law of bestowal—love thy neighbor as thyself—which becomes the tenet of all Abrahamic faiths. Even though none of the religions actually lives by this law, the very fact that they had made it the center of their faith means that people have become aware of its importance. In that, people *de facto* acknowledge Abraham's idea of love of others as a cure for humanity's ills. From this point onward, the fate of Israel and the fate of all the nations of the world will be forever entwined.

As explained before, processes that unfold in the spiritual roots manifest in their corporeal branches. For this reason, as the desire to bestow became mingled with the desire to receive on the spiritual level, the physical manifestation of that process was the spreading and mingling of the people of Israel among the nations of the world.

This does not mean that the Jews were spreading Abraham's message of love and unity to their new neighbors. The Jews did not choose to be exiled so they could spread Abraham's method. Nor did the nations who accepted them into their midst do so because they wanted to hear, much less adopt that message. Yet, because the parity-of-desires process between Israel and the nations was already underway on the spiritual level, it was happening in the physical world, as well.

Thus, by the end of the Middle Ages, the mingling of desires had reached such a stage that on the physical level it manifested in three religions whose adherents did not claim to be altruists, yet cited a fundamentally altruistic law as one of their tenets: "Love thy neighbor as thyself." Moreover, these religions—Christianity, Islam, and Judaism—were not only citing that law as their tenet, but also declared Abraham as their spiritual patriarch, hence their epithet, "Abrahamic Faiths."

In Chapter 4, we mentioned the Tokyo presentation of evolutionary biologist Elisabet Sahtouris, concerning self interest and collaboration—that every molecule, every cell, every organ and the whole body, has self-interest. And when every level shows its self-interest, it forces negotiations among the levels, which drive one's system to harmony. Indeed, even if we are oblivious to the ultimate goal of existence, subconsciously we all feel that harmony and mutual care are the only ways to create a sustainable humanity. We all have the four stages of desire within us because we are all, at the end of the day, offshoots of these four stages.

Hence, as Stage Two—the governing stage of desire during the Middle Ages—dictates, all three Abrahamic faiths adopted the commandment, "Love thy neighbor as thyself" (Lev, 19: 18) as a tenet. Thus, although "negotiations" (to use Sahtouris' term for relations) between people and nations during the Middle Ages were often far removed from what we might deem harmonious, the end result was a rather consolidated Europe with respect to religion, whose basic (declared) tenet is *altruistic*—abiding by the law of yielding self-interest, even if its realization was far less selfless.

We already know that Stage Two in the evolution of desires marks the first appearance of the desire to give *within* the desire to receive. Indeed, the commandment to love others as much as one loves oneself is in perfect congruence with Stage Two. However, our universe was created when Adam's soul broke, when its "organs" became self-centered. As a result, the law of loving others appears in our world as a commandment that one must make efforts to follow. If our nature were that of true bestowal, we would not even need this law because we would *naturally* love to bestow as much as we currently love to receive.

Yet, if our nature were one of bestowal, we would never become equal to the Creator. The most we would achieve is similarity of desires with the Creator, but we would be devoid of everything we gain by struggling with our desires. This struggle, as hard as it is, grants us unique observations. By comparing our own nature to the universal Nature, we learn the difference between giving and receiving, the knowledge that there can be giving in receiving, and the joy and fulfillment that come with being able to love. These emotions can only arise when one has experienced the *inability* to love.

But beyond all of these gifts is the greatest, uniquely human gift: freedom of choice. The difference between a mature adult and a youth in our world is in the permission, ability, and *freedom* to make one's own choices. In the spiritual realm, only humans have that ability because only humans possess both natures—that of reception and that of bestowal—assuming they have acquired that nature by following the law of yielding self-interest.

Once we obtain the nature of bestowal, we understand why it is necessary for both natures to exist within us, why we must begin with a nature of reception, acquire the nature of bestowal, and install the latter over the former out of our own free choice. Only by doing so can we truly perceive the works of Nature, with *all* her facets and subtleties. And only when we perceive all that will we be able to *consciously* live by the law of yielding self-interest in favor of the interest of Nature, because we will have achieved the *Thought of Creation*. And when we achieve that thought, we will truly become Creator-like.

The establishing of Abrahamic faiths in the hearts of millions created a first-ever bridge between inherently self-centered people and the principle of bestowal. For the first time, people felt that giving could yield them profit. Albeit this is an egocentric type of altruism, at that point in time and at that stage in the evolution of the desire to receive, this was the closest to altruism that people could come.

Thus, even though the epics and prophets differ from one faith to another, the end result is that all three Abrahamic faiths ascribe great importance to Israel, because every person whose soul has been touched by the Abrahamic tenet, "Love they neighbor as thyself," subconsciously strives for that state as it truly is in spirituality.

Today, the mingling has so expanded that the yearning for spirituality exists in virtually every person in the world. This, as Baal HaSulam explains in his essay, "The Love of the Creator and the Love of Man,"[108] is a result of the purpose of Creation, that "All the nations shall flow onto It" (Isaiah 2:2), meaning that all people will attain life's creating force. And for this to unfold, all the nations, all forms of desires in the world must be incorporated with the desire to bestow.

THE MAGNA CARTA

Since the desire to receive is a perpetually evolving force, Abrahamic faiths were not the only phenomena that evolved during the Middle Ages. Especially during the late Middle Ages, more and more people began to strive for personal emancipation and personal expression—in art, erudition, as well as economic independence.

In 1088, the first European university was established in Bologna, Italy. Then, between 1150 and 1229, universities appeared in Paris, Oxford, Cambridge, Salamanca, Montpellier, Padua, Naples, and Toulouse.

In civil law, too, the budding of momentous shifts were en route to change the face of European society. The Magna Carta Libertatum issued in 1215, and subsequently the *habeas corpus*, provided a first-ever protection of subjects, even fettered ones (albeit a limited one, at first), from the whim of the hitherto all-powerful king. And although these changes were initially applied only in England, they laid the foundations for democracy and the Age of Enlightenment through the whole of Europe.

The invention of the hourglass in the 11th century, and the compass, invented circa 1300, allowed for navigation across seas and oceans. This enabled Europeans to explore the world and to bring Christianity to such remote continents as America and Africa, thus spreading the Abrahamic tenet to even more nations.

Further assisting in the spread of knowledge and ideas was the revolutionary invention of the Gutenberg printing press in the mid 15th century. While literacy became prevalent only in the

19th century, when the price of paper became more affordable, the relative ease of printing books helped spread knowledge and ideas throughout Europe. As a result, the concepts of the Renaissance, which started in 14th century Italy, could be circulated much more quickly, which set the stage for a new era. And while the populace was still under the arbitrary rule of feudalists, many people's minds and hearts were beginning to mingle and reciprocate in an evident, almost tangible manner.

In his "Preface to the Wisdom of Kabbalah,"[7] Baal HaSulam describes how at its end, each stage prepares for the onset of the next. In much the same way, the developments and changes during the late Middle Ages mark the end of the era, as well as the beginning of the next—the Renaissance. And since, as Kabbalah explains, events in our world are triggered by the evolution of the desire to receive, these events demonstrated that the world was now ready for the next stage in the evolution of desires—Stage Three—whose onset is marked by the next seminal composition in Kabbalah, *The Tree of Life*.

8

THE RENAISSANCE AND BEYOND

Preceding every new stage in the evolution of desires, the appropriate precursor appears. First, there was Abraham; he was the Root. Then there was Moses, representing Stage One, followed by Rabbi Shimon Bar-Yochai (Rashbi), who corresponds to Stage Two. And now the time has come for Stage Three.

The emergence of Stage Three in the evolution of desires roughly corresponds to the advent of the Renaissance in Europe. Its harbinger was the greatest Kabbalist since Rashbi: Isaac Luria (the Ari)–founder of the Lurianic Kabbalah, the most systematic and structured school of Kabbalah. Today, it is the predominant teaching method, thanks to the 20th century commentaries of Baal HaSulam, who interpreted the writings and adapted them to the scientific/academic mindset of the 20th and 21st centuries.

Despite his short life, the Ari (1534-1572) produced numerous texts with the help of his prime disciple, Rav Chaim Vital. The Ari did not write his texts by himself. Instead, he would speak and Chaim Vital would write down his words.

After the Ari's early demise, Vital and several of his relatives compiled the Ari's words into cohesive texts. For this reason, many scholars have ascribed the Ari's writings to Chaim Vital and not to his teacher. Yet, even though Vital was the scribe, the provider of the information is undisputedly the Ari.

In Chapter 2, we described Stage Three as an "inverted" modus operandi, where the act is reception but the intention is to give. This was true for the initial four stages of desire. However, after the breaking of Adam's soul, the prevailing intention in the collective soul—of which we are all parts—has been inverted and regressed from bestowal to reception. And because we are all parts of Adam's soul, the hidden intention in all humans is to receive, as well. Clearly, when everyone wishes to receive, and none wish to give, it induces an unsustainable situation.

Yet, all stages appear so we may correct them. At every level of Nature, this correction occurs naturally, because the only way to sustain anything, from mineral through plant to animal, is to have all the elements contributing to the survival of the mineral, plant, or animal. Yet, in humans, as we explained in Chapter 6, this (one) sustainable state must be achieved through man's awareness. Without awareness, we go where our desires take us, and in Stage Three, they begin to take an ominous direction.

Indeed, the period from the Renaissance to the beginning of the 20th century saw two processes that fundamentally changed people's lives. One was the development of weapons, such as rifles and artillery, and the initiation of maritime discovery voyages by intrepid explorers who conquered new lands and subsequently exploited their native inhabitants and natural resources.

The other was the advent of modern science, but more than that—the "discovery" and extolling of the individual. This latter shift manifested in the thriving of art in all its forms, and most important, in the booming of humane movements such as Humanism and The Enlightenment. The Bill of Rights, the Edict of Nantes, and the Communist Manifesto are only some of the numerous changes that have laid down the basis for what we now call "the free world."

Alongside these profound transformations, Kabbalah needed its own "reformer." At the deepest level of existence, the shifts just mentioned were happening because a new level of desire had appeared, and this called for someone to "make sense" of these changes. This was the role of the Ari: to introduce the correction method for Stage Three. This is why the Ari's method is the most systematic and structured compared to all his predecessors' methods, matching the scientific, rational thinking of his time.

THE GREAT AWAKENING OF THE HUMAN SPIRIT

In the initial four stages of desire, Stage Three is special in the sense that it is the first time Creation initiates: it "decides" to receive (though just a little) in order to bestow. Thus, when Stage Three of the desire appeared in humanity, people and societies began to initiate changes in virtually every realm of life. New notions appeared and old ones reappeared, and all prospered under the wings of the Renaissance. Religion, science, technology, art, economy, politics (domestic and foreign), philosophy, and every other realm of life was scrutinized and modified, if not revolutionized.

The humane concepts behind the Magna Carta and the Habeas Corpus were being adopted throughout Europe and the United States, although they were often arbitrarily discarded in the face of financial and political interests such as colonialism and slavery. The 1689 "English Bill of Rights" or "Act Declaring the Rights and Liberties of the Subject and Settling the Succession of the Crown," further promoted the idea that every person is entitled to certain basic freedoms, including political freedom and freedom of speech. Put differently, the Bill of Rights allowed for expression of freedom of thought!

From the Kabbalistic perspective, these changes came about because the newly emerged desire of Stage Three calls for active reception of pleasure. Hence, people became more active in their search to better their lives and their aspiration for self-expression and self-determination as individuals. To realize their dreams, people began to develop new technologies, liberate politics from the shackles of feudalism, and establish the basis for modern economy.

In global politics, the stronger and wealthier countries began a fervid search for new lands in what is now known as the "Age of Discovery." Christopher Columbus, Vasco da Gama, Ferdinand Magellan, and Giovanni da Verrazano were only some of the many explorers who discovered new lands for their countries. These explorers not only discovered new lands and mapped them, but also paved the way for new trade routes, albeit for the most part this "trade" was really enslavement of indigenous peoples and exploitation of resources. But the result of the Age of Discovery was a new worldview and distant civilizations acknowlAs part of the new worldviews prompted by the Renaissance, the Catholic Church came under attack by Lutherans, Calvinists, Anglicans, and others who wished to liberalize Christianity and adapt it to their views. Liberalism and humanism flourished in the spirit of the Renaissance and strove to truly free man's thought for the first time since the golden age of Greek philosophy. Indeed, all through Europe it seemed as though the human spirit was awakening.

Supporting the new worldview were the revolutionary discoveries of Nicolaus Copernicus and Galileo Galilei that the Earth revolves around the sun, rather than the reverse, as was believed until their discoveries were made known. And when Francis Bacon established what became "the scientific method," which is still practiced today, you could safely declare that the "scientific revolution" was in full force. Institutions for promoting science, such as The Royal Society of London for the Improvement of Natural Knowledge, or simply, The Royal Society, substantiated science's hold on people's minds and imaginations, just as their hearts were moved by such cultural giants as Leonardo da Vinci, William Shakespeare, and Claudio Monteverdi.

Today, people often cite an exponential increase in the pace of changes. Essays such as Kip P. Nygren's "Emerging Technologies and Exponential Change: Implications for Army Transformation," published 2002 in Questia Online Library,[109] books such as *Living in the Environment: Principles, Connections, and Solutions* (G. Tyler Miller and Scott Spoolman),[110] or the eye-opening You Tube video, "We are living in exponential times"[111] are only three of numerous attempts to describe how fast our world is changing. But if you take into account the fundamental shift that occurred with the emergence of Stage Three in the evolution of desires, it is evident that the exponential growth has its roots deep in the concepts and innovations that first emerged during the late Middle Ages and early Renaissance.

In Chapters 3 and 5, we mentioned Item 38 in Ashlag's "Introduction to the Book of Zohar" where he writes, "The will to receive in the animate... can only generate needs and desires to the extent that they are imprinted in that creature alone."[112] The animate level that Ashlag mentions corresponds to Stage Three in the initial four stages, which manifests a heightened level of desire to receive, compared to Stage Two. At this level, the desire to receive "decides" to receive, as opposed to the automatic reception and rejection in Stages One and Two. In that sense, it is more autonomous than its predecessors. As a result, its corporeal manifestation—animals—is more active and autonomous than its preceding degree in the pyramid—plants. In much the same way, when the desire to receive in humans reached Stage Three, it prompted an increase in activity and aspiration for individual autonomy.

The beginning of the new era was promising. The zeitgeist, at least among the more fortunate in society, was one of liberation of minds and bodies, with such social upheavals as the Enlightenment, the Bill of Rights (first, the English and later the American version), Humanism, Reformation, and the Edicts of Nantes. With the added thriving of philosophy and science, it seemed as though soon everyone could enjoy the fruits of progress.

Yet, since at the bottom of all these encouraging shifts was the desire to receive pleasure in its broken, self-centered form (and to an even greater extent than ever before), Kabbalists responded to this outburst as a call for action. Kabbalists sensed that with the new possibilities that technology and science offered, as well as the heightened desire for self-expression, a new method of correction was required.

Thus, they began to declare that it was time to come out and show the world the long-hidden wisdom of *The Zohar*. Without it, they proclaimed, the world would not see a positive conclusion at the end of the new era. In the words of The Vilna Gaon (GRA), which numerous Kabbalists echoed, "Redemption [from egoism] depends on the study of Kabbalah."[113]

REMOVING
THE VEIL OF SECRECY

In tune with the shifts that took place at the onset of the Renaissance, Kabbalists began to remove the veil from the wisdom of Kabbalah, or at least to speak in favor of removing it. Since the writing of *The Book of Zohar*, Kabbalists have set up various obstacles before those who wished to study. It began with Rashbi's concealment of *The Zohar* and continued with declaring all sorts of prerequisites that one had to meet before receiving permission to study. The Mishnah, for instance, gives the apparently paradoxical instruction to avoid teaching Kabbalah to students who are not already wise and understand with their own mind, but the text does not specify how is one to come by wisdom if one is not permitted to study.[114]

In the Babylonian Talmud, there is a well known allegory about four men who went into a *PARDES* (an acronym for all forms of spiritual study—*Peshat* (literal), *Remez* (Implied), *Derush* (interpretations), and the highest level being *Sod*, Kabbalah). Of the four, one died, one lost his sanity, one became heretical, and only one, Rabbi Akiva, who was a giant among Kabbalists—entered in peace and departed in peace. There are other deeper and more accurate explanations to this allegory, but the story was nonetheless used to intimidate and deter people from studying Kabbalah.[115]

Another prerequisite that Kabbalists set up was to "fill one's belly with" (be proficient in) Mishnah and Gemarah before one approaches the study of Kabbalah. To justify that condition, they cited the Babylonian Talmud, which warns that one must spend a third of one's life studying the Bible, another third studying Mishnah, and the remaining third studying The Talmud.[116]

This, of course, leaves no time to study Kabbalah, so when the time came for Kabbalists to permit the study, they had to "make room" in the day for the study of Kabbalah. Thus, Kabbalists such as Tzvi Hirsh of Zidichov, "detoured" the prohibition by declaring that every day, one must "fill one's belly with" Mishnah and Gemarah, and then study Kabbalah.

There are numerous examples for Kabbalists' proclamations that Kabbalah is the means for salvation (correction of the soul, meaning to give the desire to receive the aim to bestow), and that it should not be neglected. Moreover, as a rule, the more recent the Kabbalist, the greater the tendency to give preference to the study of Kabbalah over any other form of study.

The Book of Zohar says, "At the end of days, when your composition [*The Book of Zohar*] appears below, because of it, you will set the land free [liberate the desire from egoism, i.e. correct it]."[117]

To Kabbalists, the appearance of such a systematic and structured method as the Ari's marked the beginning of the end of days, or what they refer to as "the last generation."

In his introduction to *The Tree of Life*, Chaim Vital wrote, "Even in this, last generation, we are not disgusted and we do not loathe breaching His [the Creator's] covenant with us."[118] In other words, in Vital's view, which he repeats several times in this introduction, we are in the last generation, yet we still have no wish for correction from egoism to altruism.

Moreover, he continues, "When it so happens that the days of Messiah* draw near [toward the end of correction], even little children will understand the great secrets of the wisdom. Also, it has been explained that thus far, the words of the wisdom of *The Zohar* were hidden, but in the last generation this wisdom will emerge and become known."[119]

Vital also explains that all the problems of *Adam ha Rishon*— the collective soul that we all comprise—stem from not knowing Kabbalah. In his words, "It was explained... that the sin of *Adam ha Rishon* [though Kabbalists refer to his sin as a "mistake," not as a deliberately malicious act] was that he did not choose to engage in the Tree of Life, which is the wisdom of Kabbalah."[120]

In the rest of the above quoted text, Chaim Vital attempts to ease people's approach to Kabbalah by clearing up the prevalent misconceptions that Kabbalists have been fostering since the concealment of *The Book of Zohar*. In his words, "This, itself, is the sin of the mixed multitude [a reference to those among the Jews who prohibit the study of Kabbalah], who say onto Moses, 'You speak to us...and let God not speak to us lest we die in the secrets of Torah [a common epithet to the wisdom of Kabbalah].' It is as the erroneous ones believe and say that any person who

---

* The force that pulls out of egoism. In Kabbalah, the term, "Messiah," refers to the Hebrew word, Moshech (pulling), which refers to a force that pulls one from egoism to altruism, thus correcting one's soul. The terms "redemption" and "deliverance" are also code names for this shift from egoism to altruism. Also, the coming of the Messiah refers to the time when this will happen to the whole of humanity.

engages in it [Kabbalah] will live a short life. Today, it is they who slander and give a bad name to the wisdom of truth [another epithet for Kabbalah]."[121]

In another place, he adds, "Thus far, the words of the wisdom of *The Zohar* were hidden, but in the last generation [which Vital defines as his generation] this wisdom will appear and become known, and they will study and comprehend the secrets of Torah [Kabbalah], which the former ones did not attain. By that, the objection of the fools who say, 'If the former ones did not know it, how will we?' shall be revoked. As it is explained, in these last generations they will be nourished by that composition [*The Book of Zohar*] and that wisdom will appear to them."[122]

While under the patronage of his mentor, the Ari, Chaim Vital had the privilege of learning from the highest Kabbalah authority of his time. Still, Vital was not the only voice in and around his generation to laud the need to publicize Kabbalah. Kabbalist Avraham Ben Mordechai Azulai (1570-1644), clearly expressed the need to publicize Kabbalah from his time forth: "I have seen it written that the prohibition... to refrain from open study in the wisdom of truth was only... until the end of 1490. But from then on the prohibition has been lifted and permission was granted to engage in *The Book of Zohar*. And from the year 1540, it has been a great *Mitzva* [commandment, but also good deed] for the masses to study, old and young... And since the Messiah will come because of that, and for no other reason, we must not be negligent."[123]

In the 16th century, the town of Safed, in today's Northern Israel, was the "capital" of Kabbalah. This was also the town where the Ari had lived and taught his students. The greatest Kabbalist in Safed until the arrival of the Ari was Moshe Cordovero (1522-1570), known as "the Ramak." He preceded the Ari by a few years, but he could already sense the approaching of a new stage of desire. In his book, *Know the God of Your Father*, he wrote, "The whole Torah speaks of nothing but the existence of the

Maker and His merit in His *Sefirot* and His operations in them. And the more one studies its secrets [Kabbalah], the better, since one utters His merit and does wonders in the *Sefirot*."[124]

Over time, Kabbalists sensed an increasing urgency for people to study Kabbalah because they feared that problems and calamities would ensue if people did not know life's basic modus operandi. They even began to write in favor of teaching children. Yitzhak Yehuda Sarfin from Komarno (1806-1874), for example, wrote in his book, *Notzer Hesed* (*Keeping Mercy*), "Had my people heeded me in this generation... they would have studied *The Book of Zohar* and the *Tikkunim* (corrections, part of *The Zohar*), and contemplated them with nine-year-old children."[125]

Similarly, Kabbalist Rav Shabtai Ben Yaakov Yitzhak Lifshitz (c. 1845-1910), wrote in his book, *Segulot Israel* (*The Virtue of Israel*), "May they begin to teach the holy *Book of Zohar* to children when they are still small, ages nine or ten, as it was written by the great Kabbalist... and redemption [complete correction] would certainly soon follow."[126]

To some extent, the Kabbalists succeeded in their efforts. The *Hassidut* (Hassidism) movement, established in the 18th century Polish-Lithuanian Commonwealth (today's Ukraine) by Rabbi Israel ben Eliezer, (1698–1760), known as The Baal Shem Tov (Owner of the Good Name), produced a great many Kabbalists. Once the students of The Baal Shem Tov attained sufficient proficiency in Kabbalah and a clear enough perception of the spiritual world, he sent them to other towns to continue spreading the wisdom. The students of The Baal Shem Tov nurtured more students, helped them attain spiritual perception, too, and in turn, sent them on their way to further spread the wisdom. Thus, a vast movement was formed, whose heads were all Kabbalists.

Yet, in time, just as it happened to the people of Israel before the ruin of the Second Temple, the spiritual level of the teachers declined until they lost their spiritual attainment altogether. Even so, the positive effects of the *Hassidut* cannot be overrated when considering The Baal Sham Tov's success in introducing the hitherto hidden wisdom to the masses.

KABBALAH'S REACH EXTENDED

Although Kabbalah had been a secret wisdom for over a millennium, Kabbalistic texts could always be found if one truly wished to study them. During the Renaissance, many scholars not only discovered Kabbalah books, but apparently studied them enthusiastically and treated Kabbalah as a wisdom of great merit.

In the previous chapter, we mentioned Johannes Reuchlin (1455-1522), who claimed that Pythagoras received his knowledge from the Jews, i.e., Kabbalists, and that the term philosophy appeared when Pythagoras translated the word, "Kabbalah," into the Greek name, "philosophy." But Reuchlin was not the only one. Many acclaimed scientists and thinkers spoke favorably of Kabbalah and urged their readers to explore it, striving to clear the misconceptions and stigmas surrounding it.

One of the most notable philosophers who showed a keen interest in Kabbalah was acclaimed playwright, author, and scientist, Johann Wolfgang von Goethe (1749-1832). In *Materialien zur Geschichte der Farbenlehre*, Goethe wrote, "The whole chorus [assembly] of those who gathered—Jews, Christians, pagans and holy men, Fathers of the Church and heretics, Councils [Synod] and popes, reformers and opponents altogether, while they explain... [they] do this by the way of Plato or Aristotle, consciously or unconsciously, as... the Talmud and Kabbalistic treatment of the bible convince us."[127]

CONNECTING AND COMMUNICATING

The earlier centuries of Stage Three in the evolution of desires provided the basis for expansion of land and ideas. The Age of Discovery, the Scientific Revolution, Humanism, the Reformation, and the Enlightenment movement were all parts of a profound change that opened people's minds and expanded their worldviews. These movements and ideologies enabled people to explore beyond their childhood rearing and reflect on life and its meaning. The Romantic period in classical music, the *Sturm und Drang* (storm and stress) literary movement, and the Impressionist style of painting underscored the emphasis on personal experiences and emotions in art, and in fact, presented a trend that would only strengthen in the 20th century. This trend, which eventually produced the narcissism epidemic that Twenge and Campbell refer to (see the Introduction and Chapter 5), was a forerunner of Stage Four in the evolution of desires.

But the existence of such noble ideas as equal opportunities, human rights, and freedom of speech was not enough to set off a new era. To do that, there had to be means to communicate these ideas. The 18th and especially the 19th centuries facilitated precisely that—mass communication and mass transportation.

The steam engine, first invented in the 17th century, was improved dramatically in the subsequent two centuries, and became a primary provider of motive power (engine) for industry and transportation. Toward the end of the 18th century, steam engines began to be used in boats. In the next century, these engines had so improved that they became the primary source of motive power in boats and ships.

On land, the steam locomotive changed the face of 19th century transportation. The first attempts to develop a steam engine locomotive date back to the second half of the 18th century. However, it was only upon the advent of George and Robert Stephenson's 1829 multi-tube boiler, Rocket, that a commercially viable steam locomotive was built. In fact, the Rocket locomotive was so successful that improved versions of it were in commercial use deep into the 20th century, and even the beginning of the 21st (Image no. 8). And although it has become a rare sight, steam engines are still in use today in locomotives. Thus, with such an efficient means of transit, commuting became easy and migration of people far more frequent.

Image no. 8: Brand new steam locomotive
60163 Tornado, manufactured in England, 2008.

Private transportation, too, was developing around the same time. Various forms of "horseless carriages," as automobiles were called, have existed since the end of the 18th century. But until the last quarter of the 19th century, they were treated as bizarre, and often a nuisance. In 1865, the Locomotive Act in Britain restricted the speed of horseless vehicles to 4mph in open country and 2 mph in towns. Furthermore, the Act required *three* drivers for each vehicle—two to travel in the vehicle and one to *walk ahead* waving a red flag.

But in 1876, Nikolaus August Otto invented a successful four-stroke engine, known as the "Otto cycle," and that same year the first successful two-stroke engine was invented by Scottish engineer, Sir Dugald Clerk. Ten years later, the first vehicles using internal combustion engines were developed at roughly the same time by two engineers working in separate parts of Germany—Gottlieb Daimler and Karl Benz. They simultaneously formulated highly successful and practically powered vehicles that worked much like the cars we use today. This was the start of The Age of Motor Cars.

Early in the 20th century, the final terrestrial frontier was conquered—the sky. According to the Smithsonian National Air and Space Museum, "On December 17, 1903, at Kitty Hawk, North Carolina, the Wright Flyer [Orville] became the first powered, heavier-than-air machine to achieve controlled, sustained flight with a pilot aboard."[128] From then on, even the sky was not off limits for humanity.

In the time frame between the writing of *The Tree of Life* and the beginning of the 20th century, our desire to govern and to profit led us to develop such capabilities in science, technology, communication, and transportation that by the start of the 20th century, all major land masses were known, connected, and regularly trading with one another. Thus, the world had effectively become a single entity, a global village. And while this may not have been evident to ordinary citizens at the time, the 20th century, with its joys and sorrows, would thoroughly demonstrate our connectedness and interdependence.

As we said at the beginning of this chapter, preceding every new stage in the evolution of desires appears the appropriate precursor. In the case of Stage Four, its precursor was not just a Kabbalist who could explain matters better than any of his predecessors, but almost an entire century served as a forerunner of a new era. The 20th century not only foretold, but even facilitated the advent of the new desire. For this reason, the 20th century merits an entire chapter of coverage.

9

ONE WORLD

On the face of it, the 20th century seems like the beginning of a new stage in the evolution of desires. Every single realm of human engagement was revolutionized (and often re- or counter revolutionized) during this century. Indeed, the pace of change during this century has so increased, life has begun to change at an exponential pace.

But even more astonishing than the pace of progress was the pace of globalization. The process of becoming a single economic system that began with the Age of Discovery and colonialism culminated in the 20th century. At the century's end, virtually no country remained completely self-sufficient.

Although the rapid expansion and change in all realms of life is clearly evident, its scope and speed are so alarming that in my view, it is worth a short reflection. However, if you feel reviewing some of the major developments during the 20th century is unnecessary, you are welcome to skip right to the next sectIn the year 1900, the world population was approximately 1.6 billion. By the end of the century, it was in excess of six billion. In 1900, the average top speed of a car was seven mph. A hundred years later, even typical family cars could reach 130 mph. Moreover, the primary means of transport had changed from carriages, bicycles, and walking to driving. By the turn of the 20th century, the majority of walking was done on treadmills at home, in parks, or in fitness gyms, and the same could be said for cycling.

For overseas journeys, jetliners have completely replaced passenger ships, and travel time between continents had dropped from several weeks to several hours (albeit for shipment of goods, the primary means of transport is still cargo ships rather than planes). And (quite literally) above all, to help ships and cars navigate, to alert them of bad weather, and to survey enemy territory, we have positioned satellites in space.

With respect to technology, life has changed not only in how fast and how comfortably we travel, but also in the instruments we use in our daily lives. Such devices as telephones (and later cellular phones), light bulbs, radios, televisions, and computers were either unheard of or were just making their debut in the early 1900s. At home, life has never been easier. Washing machines, clothes dryers, refrigerators, freezers, vacuum cleaners, electric stoves, and (since the 1970s) microwave ovens, all have become household appliances.

In 1900, the popular entertainment was vaudeville (a traveling circuit of live acts featuring magicians, acrobats, comedians, trained animals, singers, and dancers), as well as silent black-and-white films and ragtime. By the year 2000, "canned" films were in full color and Dolby surround stereo sound, and professional sports had become a major entertainment outlet. Music offered countless styles, each with its own numerous sub-styles: rock, folk, blues, classical, jazz, pop, hip-hop, trance, ethnic music of all sorts, and the list is endless. But not only music, dance, theatre, visual arts, photography, and every other form of art have exponentially expanded in diversity. Computer games, too, have become very popular by the end of the 20th century, and the internet was beginning to expand its presence in people's homes. Additionally, people no longer needed to leave their doorsteps for entertainment or to gather information because they had radios, televisions, record/cassette/CD players, and VCRs or DVDs.

Alas, the technological advances of the 20th century were (and still are) used detrimentally with devastating results: war, occupation, oppression, and tyranny became exponentially more effective and destructive, resulting in two world wars and several genocides within the time frame of a single century.

The two world wars changed the world map dramatically and ended the age of colonialism (with some exceptions such as India, which gained independence from England in 1947, or Algeria and other nations under French rule, which gained their own in the 1950s and 1960s). This allowed numerous new countries to experience independence for the first time, though the gap in wages, infrastructure, and standard of living between the powerful post-empires and the newly liberated countries not only remained, but even widened.

In the 20th century, science had drastically changed the way we view the world. Einstein's Special and General Theory of Relativity, followed by the advent of quantum mechanics, have revolutionized the way scientists perceive the world, paving the way for numerous innovations from lasers to microprocessors and everything derived from them. Genetics was significantly developed, the structure of DNA was determined, and by the turn of the century, the first mammal, Dolly, the sheep, was cloned.

In astronomy, the Big Bang theory was proposed and the age of the universe was determined at roughly 14 billion years. Also, our observation capabilities have been dramatically improved with the 1990 launch of the Hubble Space Telescope.

And the last, but certainly not least topic on the list is medicine and health. According to a December 28, 2007 National Vital Statistics Report, written by Elizabeth Arias, PhD, for the Centers for Disease Control and Prevention (CDC), an American male Caucasian infant born in 1900 could expect to reach the age of 46 (32 if he were Afro-American). In 2000, the numbers were 74 and 68, respectively.[129] This was made possible by improved medical hygiene such as sterilization of instruments used for surgery and the use of protective clothing by medical personnel, and improved personal hygiene, such as hand washing, as well asa host of vaccines that were developed along with the rapid spread of antibiotic medications.

Also, technological advances made X-rays a powerful diagnostic tool for a wide spectrum of diseases, from bone fractures to cancer. In the 1960s, Computerized Tomography (CT) was invented, and a decade later, Magnetic Resonance Imaging (MRI) was developed.

All these and many more 20th century innovations and shifts made the past century a landmark of unique position in history.

INVISIBLE LINKS

On at least three accounts, the world witnessed the effect of the invisible links that tie us into a single system. In the two World Wars, virtually entire continents engaged in active fighting. The Great Depression sent multiple financial tsunami waves across the globe, destroying the lives and incomes of millions. According to the Encyclopedia Britannica, "Since the U.S. was the major creditor and financier of postwar [WWI] Europe, the U.S. financial breakdown precipitated economic failures around the world... Isolationism spread as nations sought to protect domestic production by imposing tariffs and quotas, ultimately reducing the value of international trade by more than half by 1932."[130]

Yet, despite the evidence, humanity did not recognize that it was a closed, interdependent system. Each time adversities unfolded, countries turned to protectionism and isolation by raising tariffs, employing punitive measures against the apparent wrongdoers, and ignoring or overlooking the fact that adversities are never created or executed by a single culprit. Rather, they have always been the culmination of a prolonged process that involved many partakers.

Therefore, when you realize how deeply we are all connected, that at the deepest level, we are actually a single entity, it becomes very hard to point a blaming finger at any one perpetrator. At that point, you begin to examine issues and situations from a broader perspective, understanding that what each of us does affects every person in the world. But for this, one must be aware that all people form a single soul (desire to receive), whose self-centered *modus operandi* blinds its parts to the truth of their interconnectedness and interdependence.

For as long as humanity was evolving under the influence of desires Zero through Two, our obliviousness to our interconnectedness was tolerable. In Stage Zero, there was basically no discernible desire to receive; man was a part of Nature. In Stage One, during Abraham's time, egoism appeared for the first time. Yet, at that point, humanity was in its infancy and there was no danger of us causing irreversible harm to ourselves or to the environment. In Stage Two, there was evidently more egoism but it, too, was managed, primarily by religion, as we have shown in Chapter 7.

In Stage Three, the desire to receive became active. As a result, since the debut of Stage Three in the late Middle Ages, humanity has launched a frenzy of accelerating development and growth which have now reached an uncontrollable rate. As we will see below, this rate of growth has long been recognized by science, as well as by Kabbalah.

In the previous chapter, we cited researchers' observations that humanity is advancing at an exponential rate. But perhaps the most compelling evidence of science's recognition of this trend is that of Charles Darwin. Through his observations and those of his predecessors, we learn that exponential growth is not a recent phenomenon. Instead, exponential growth is how the *whole* of Nature works.

In *On the Origin of the Species*, Darwin discusses exponential growth and quotes Swedish botanist, Carolus Linnaeus (1707-1778), who also observed this pattern: "There is no exception to the rule that every organic being increases at so high a rate, that if not destroyed, the earth would soon be covered by the progeny of a single pair. Even slow-breeding man has doubled in twenty-five years, and at this rate, in a few thousand years, there would literally not be standing room for his progeny. Linnaeus has calculated that if an annual plant produced only two seeds— and there is no plant so unproductive as this—and their seedlings next year produced two, and so on, then in twenty years there would be a million plants"[131] (*On the Origin of the Species*, The Struggle For Existence, pp 117-119).

When desires are small, such as in plants, animals, or even in the early stages of the evolution of desires in humans, Nature finds ways to balance the exponential growth rate by presenting equally powerful elements such as competing plants and animals that form a delicate balance. This is why Darwin writes in the above quote, "Every organic being increases at so high a rate, that *if not destroyed*, the earth would soon be covered..."[132]

Put differently, Nature's own mechanisms guarantee that excess reproduction of plants and animals would be restrained.

But when desires grow exponentially in a dominant species, and especially when they manifest a self-centered trend such as begun to manifest in Stage Three, the environmental balance is breached and a serious problem arises.

THE EXPONENTIAL EFFECT

To better understand the shift that unfolded during the 20th century and is unfolding still, we need to understand the nature of exponential growth. The decisive factor in exponential growth is not the initial quantity, but what is known as the "doubling time speed." This refers to the length of time it takes for the amount of the measured object to double.

To understand the difference between exponential growth and linear growth, consider the following scenario: Mrs. A is a poor woman with only one dollar in her savings account. Mr. B, on the other hand, is much better off and has 10,000 dollars in his account. Both Mrs. A and Mr. B save what they can for a rainy day, and have thirty years of work ahead of them before they retire and collect their pension.

Mrs. A's savings grow exponentially and her doubling time is one year. Thus, after one year she has two dollars in her account ($\$1 \times 2^{1(year)} = \$2$); after two years she has four dollars ($\$1 \times 2^{2(years)} = \$4$), and after three years she still has a meager eight dollars in her account ($\$4 \times 2^{3(years)} = \$8$).

Mr. B's savings grow linearly, hence adding a fair 10,000 dollars to his account each year.

After five years, it seems as though Mrs. A is destined to a life of indigence, with only thirty-two dollars in her account, while Mr. B seems headed for a life of relative affluence with as much as 50,000 dollars in the bank. However, if they continue their savings curve for the entire thirty years until their retirement, at the end of the term, Mr. B will have accumulated a handsome sum of $10,000 x 30 years = $300,000 in his savings account.

Mrs. A, on the other hand, will be no longer poor. After thirty years of exponential saving, her account will have accumulated a whopping 1,073,741,824$ ($1 x 2^{30})—over a *billion* dollars!

As we said above, Kabbalists have long known about the exponential pattern of growth in human nature. They described it in a frequently cited quote from the 1,500-year-old text, *Midrash Rabbah*: "If one has 100, he wishes to make them 200, and if he has 200, he wishes to make them 400."[133]

Yet, there is a not-so-subtle difference between the exponential doubling time in an ordinary exponential formula and the Kabbalistic doubling time. In a traditional exponential formula, the doubling time is fixed. When the annual growth of a country's GDP, for example, is seven percent, the doubling time for the GDP is ten years. Thus, economists can plan ahead even when the growth is fast, since it is still predictable and therefore somewhat manageable.

The growth of desires, however, is unpredictable. In desires, as the above quote demonstrates, what doubles the desire is not a fixed length of time, but the fact that one has *satisfied one's wish*. Notice that the quote says, "If one has 100, he wishes to make them 200," etc. This means that the condition for acquiring a twice-as-strong desire is the realization of the former one. In other words, *you can never have what you want because the minute you have it, you want twice as much*.

Thus, if Mrs. A's desire to have one dollar in her account were satisfied, she would immediately want to have two dollars. And as soon as she had two, she would immediately want four dollars in her account. Hence, the Kabbalistic exponential formula dictates that Mrs. A's desires would always double one step *ahead* of her accomplishments. Consequently, as her accomplishments doubled, so would her desires, leaving her not only with a sense of eternal deficiency, but one that doubled every time she obtained what she wanted.

If Mr. B's wish was to save $10,000 each year, then for the past thirty years he has been a content (if not happy) man and can now retire in peace. Mrs. A, however, whose initial wish was for only one more dollar, has now become deficient of over a billion dollars, because this is what she has in her account.

What's more, with the exponential growth of her wealth (and consequent dearth), she is destined to a life of hopeless pursuit of wealth and happiness, which will only lead to misery and pain for the rest of her life. The Babylonian Talmud says of one with this type of desire, "Anyone who is greater than his friend [in this case financially], his desire is greater than himself,"[134] and (*Midrash Rabbah*, as earlier cited), "One does not leave the world with half of one's wish in one's hand."[135]

THE WORLD WIDE WEB

As we have just noted, human desires double every time we satisfy them. This forces us to continuously innovate, devise new instruments, explore new seas, and conceive new ideas in order to obtain what we want. During Stage Three in the evolution of desires, when desires first became active, the effects of the exponential pattern were clearly showing in the accelerated pace of progress.

Thus, in search of new avenues for pleasure, we have turned the world into a web of trade routes by air and by sea, and by numerous technologies of communication. The World Wide Web is not just a virtual entity that lives in our computers; it is a name that describes the reality of our lives. This was recognized many years ago by sociologists, as well as by Kabbalists.

Today, globalization and financial interdependence are well-recognized facts. Yet, globalization is far more than financial interdependence; it entails a profound mingling of culture, society, civilization as a whole, and in the end, a common fate. Professor of International Relations and prolific author on globalization, Anthony McGrew, made very clear statements about the impact of this process on human society. In an essay titled, "A Global Society?" he writes, "In comparison with previous historical epochs, the modern era has supported a progressive globalization of human affairs. The primary institutions of western modernity—industrialism, capitalism, and the nation-state—have acquired, throughout the twentieth century, a truly global reach. But this has not been achieved without enormous human cost... While early phases of globalization brought about the physical unification of the world, more recent phases have remade the world into a single global system in which previously distinct historical societies or civilizations have been thrust together. This... defines a far more complex condition, one in which patterns of human interaction, interconnectedness, and awareness are reconstituting the world as a single social space."[136]

Kabbalist Yehuda Ashlag, too, recognized the trend and its hazards, and explained it from the perspective of the evolution of desires. In his essay, "Peace in the World," Ashlag provides both his observation of the state of the world in his time, as well as the approach humanity should adopt if it is to cope with the situation. In the essay, he writes, "We have already come to such a degree that the whole world is considered one collective and one society. This means that because each person in the world draws life's marrow and livelihood from all the people in the world, one is coerced to serve and to care for the well-being of the whole world."[137]

Subsequently, Ashlag explains how we are all connected and interdependent, and concludes as follows: "Therefore, the possibility of making good, happy, and peaceful conducts in one state is inconceivable when it is not so in all the countries in the world, and vice versa. In our time [he wrote the essay in 1934], the countries are all linked in the satisfaction of their needs of life, as individuals were to their families in earlier times. Therefore, we can no longer speak or deal with conducts that guarantee the well-being of a single country or a single nation, but only with the well-being of the entire world because the benefit or harm of each and every person in the world depends and is measured by the benefit of all the people in the world."[138]

In the last paragraph of that section, Ashlag predicts that mere intellectual, scholastic comprehension of the situation will not be enough for people to internalize their interdependence. Rather, life's experiences will compel them to do so: "And although this [interdependence] is in fact known and felt, the people in the world have not yet grasped it properly... because such is the conduct of development in nature, that the act [impact of interdependence on our lives] comes before the understanding, and only actions will prove and push humanity forward."[139]

In hindsight, we can say that regrettably, Ashlag's prediction came true on more than one occasion in the 20th century, and in the most gruesome of ways. In "Peace in the World," as well as in several other essays, Ashlag predicts what will happen if we continue to let the act precede the understanding. He suggests how we should conduct ourselves in order to build a sustainable, and indeed desirable existence. Now that we understand our interdependence, these suggestions will be the topic of discussion through the rest of the book.

10

THE AGE OF FREE CHOICE

In Chapter 6, we said that unlike all other elements in Nature, human beings have the power to change the environment. This gives us something that no other creature has: freedom of choice. Put differently, human beings can choose to be like the Creator—giving—and acquire the power and cognizance that come with it, by adopting the law of yielding self-interest before the interest of the environment. Or they can remain as they were born—self-centered, with limited understanding of Nature, and paying the price for their errant ways throughout history. But to choose to be like the Creator, which, as we said in Chapter 1, is synonymous with Nature, people must know what the term, "Creator," means and how they can become like it.

We also said (Chapter 3) that the whole of reality consists of a single, broken entity, called "Adam's broken soul" or "the broken soul," and that the term, "soul," refers to a desire to receive with an intention to bestow. When Kabbalists say that something is broken, they are not referring to any physical shattering, but to the tearing of the links between all parts of the soul, the collective desire that constitutes our reality. This tearing occurs when the pieces in the soul begin to operate in their own interest rather than in the interest of the system. It is as if cells in an organism begin to operate for themselves, causing the organism to die and disintegrate.

Yet, unlike organisms, the soul cannot disintegrate because it is a single desire. So while the links are there, we can enjoy the benefits of the connection. Healthy cells benefit from each other in an organism, supporting each other's existence, but cancer cells compete with each other for blood and nourishment, thus constantly harming each other. In the case of humanity, we are not even *aware* that we are connected, which prevents us from trying to connect iBut regardless of our awareness, we are very much connected. On September 10, 2009, *The New York Times* published a story titled, "Are Your Friends Making You Fat?" by Clive Thompson.[140] In his story, Thompson describes a fascinating experiment performed in Framingham, Massachussets. In the experiment, certain details of the lives of some 15,000 people were documented and registered periodically over more than fifty years. This allowed researchers Dr. Nicholas Christakis, a medical doctor and sociologist at Harvard, and James Fowler, at the time a Harvard political science graduate student, to create a map of interconnections and examine the long-term impact that people had on one another.

Christakis and Fowler established that there was a network of interrelations among more than 5,000 of the participants. Christakis and Fowler discovered that in the network, people affected each other and were affected by each other. These effects seemed to work not just in social issues, but surprisingly, with physical issues, as well.

"By analyzing the Framingham data," Thompson wrote, "Christakis and Fowler say they have for the first time found some solid basis for a potentially powerful theory in epidemiology: that good behaviors—like quitting smoking or staying slender or being happy—pass from friend to friend almost as if they were contagious viruses. The Framingham participants, the data suggested, influenced one another's health just by socializing. And the same was true of bad behaviors—clusters of friends appeared to 'infect' each other with obesity, unhappiness, and smoking. Staying healthy isn't just a matter of your genes and your diet, it seems. Good health is also a product, in part, of your sheer proximity to other healthy people."[141]

Even more surprising was the researchers' discovery that these infections could "jump" across connections. They explained that people could affect each other even if they did not know each other. Moreover, Christakis and Fowler found evidence of these effects even three degrees apart (friend of a friend of a friend). In Thompson's words, "When a Framingham resident became obese, his or her friends were 57 percent more likely to become obese, too. Even more astonishing... it appeared to skip links. A Framingham resident was roughly 20 percent more likely to become obese if the friend of a friend became obese—even if the connecting friend didn't put on a single pound. Indeed, a person's risk of obesity went up about 10 percent even if a friend of a friend of a friend gained weight."[142]

As Christakis and Fowler described in *Connected: The Surprising Power of Our Social Networks and How They Shape Our Lives*, their [then to be published and now celebrated] book on their findings: "You may not know him personally, but your friend's husband's co-worker can make you fat. And your sister's friend's boyfriend can make you thin."[143]

Quoting Christakis, Thompson wrote, "In some sense we can begin to understand human emotions like happiness the way we might study the stampeding of buffalo. You don't ask an individual buffalo, 'Why are you running to the left?' The answer is that the whole herd is running to the left."[144]

Similarly, in his essay, "The Freedom," Baal HaSulam writes, "He who strives to continually choose a better environment is worthy of praise and reward. But here, too, it is not because of his good thoughts and deeds, which come to him without his choice, but because of his effort to acquire a good environment, which brings him these good thoughts and deeds."[145]

Thus, while the links themselves exist, as the above study demonstrates, our self-centeredness prevents us from being aware of them. "Christakis and Fowler's strangest finding," writes Thompson, "is the idea that a behavior can skip links—spreading to a friend of a friend without affecting the person who connects them. If the people in the middle of a chain are somehow passing along a social contagion, it doesn't make sense, on the face of it, that they wouldn't be affected, too. The two researchers say they don't know for sure how the link-jumping works."[146]

Indeed, we act as if we are not connected, when we are actually very much so. Today, our interconnectedness has become interdependence; therefore, the gulf between reality and our incessant denial of it is posing a real threat. This is the real cause of the worldwide crises we have been experiencing.

MANDATORY FREE CHOICE

On the lower levels of desire—in Stages One through Three— Nature mends the ties described in the previous section by itself. In the process of evolution, the elements in Nature that follow the rule of yielding self-interest before the interest of their host system survive and form the basis for the next level in evolution. The ones that do not yield their self-interests perish.

Thus, gradually, Nature built the universe, galaxies, our solar system, and planet Earth. Then, layer by layer, as we have shown in Chapter 4, life on Earth was formed.

As biologist Sahtouris so eloquently explained, initially each new creature conducts itself selfishly, oblivious to the existence and needs of other creatures. But the struggle among the creatures forces them, as she put it, to "negotiate," eventually leading to the creation of homeostasis—the stability necessary for the persistence of life.

In this manner, life on Earth evolved stage by stage until at Stage Four in the evolution of desires, Homo sapiens appeared. Initially, humans were just like all other creatures. Just as desires evolve in the whole of Nature, our desires, too, evolved stage by stage, from Zero through Four. In Stages Zero through Two, the desires for greed, control, and cognizance were not potent enough to separate us from nature to a point that threatens our existence. Like all other elements of Nature, we were forced to negotiate and accept the power of the elements as one of life's necessities. However, history shows we were not quite as pliable and tolerant toward other humans.

But as described in Chapter 8, roughly since the 15th century, Stage Three took the hold. Since then, cravings for self-expression and personal excellence have been growing in us and expanding exponentially.

There is a peculiar quality to the desires for recognition and personal distinction. Although these desires reflect a self-centered nature, since they aim to present the individual who possesses them as superior to others, they also compel those who have them to connect to others. This is so because to be superior to others, one must measure one's qualities, achievements, efforts, and possessions compared to those of others. If I do not compare myself to others, over whom can I be superior?

Thus, superiority *dictates* comparison, and hence forces the egocentric human of Stage Four to perpetuate connections with others. And the more egocentric we are, the more we want to feel superior to others, and are thus forced to tighten our connections to others.

In fact, the very word, "egocentric," implies that there can also be another center to our thoughts. And the negative stigma attached to egoism implies that we instinctively know which direction is best for us—altruism, being "other-centric."

The question is, "Why are we not acting like the rest of Nature, in the way that seems to be in our own best interest, as well?" The answer is that it seems it would be best if everyone were altruistic, but (save for very few) because of our egos, we want everybody else to go first. We all agree with the idea of altruism, but are paralyzed when it comes to executing it. Until we see that everyone else is doing it and know for certain that we will not lose by giving, we cannot give.

As a result, altruism does not seem like a good idea, but like a naïve one, even dangerous—if I were to go first and then were exploited. In consequence, Nature's way, which seems like the right way for us, actually appears to be the wrong way in practice. This is why it seems unreasonable that we should choose it.

But at the same time, as we have shown throughout this book, *only* the altruists survive. We are already connected, and we already affect one another, therefore harming each other with our treacherous intentions toward others. Put differently, our egoism is already taking its toll on us, so as we can see, the choice of altruism is both mandatory and utterly unappealing.

Yet, it is this unattractiveness that makes it a free choice. If it were appealing, we would do it automatically, following our egocentric intentions, and it would no longer be altruism, but disguised egoism all over again, which would lead to our eventual destruction.

But there is another reason why free choice is a must for us humans. In the beginning of the book, we said that according to Kabbalah, the purpose of Creation is to become like the Creator, just as the purpose of a child is to become a grownup, like its parents. And just as a child must learn to make choices freely on issues that concern corporeal life, Creation, meaning we, must learn to make choices freely in regard to spiritual life.

When Kabbalists refer to the spiritual life, they refer to choices whether to act out of egocentric motives or out of socio-centric or Creator-centric motives. In choosing to be socio-centric, one attains the purpose of one's existence of becoming Creator-like, with all the capabilities and responsibilities that go with it.

In Chapter 2, we mentioned Meltzoff and Prinz's book, *Perspectives on Imitation*, where they describe the importance of imitation and identification with role models in rearing children. Yet, not only children learn this way; it is how we *all* learn. If we were not affected by each other's wishes and behaviors, fashion would have been impossible, since no one would follow anyone else. Moreover, we would not progress, since nothing in our neighbors would evoke our envy and drive us to improve our own lives. This would halt the wheels of progress instantaneously. By performing acts of altruism we imitate the Creator—the life-giving force that creates and propels everything that happens. And just as children learn how to be grownups by imitating them, we will learn how to be like the Creator by imitating it, as well.

It could be argued with a great degree of merit that many people perform acts of altruism, yet none of them seems to have acquired the qualities or capabilities of the Creator. Indeed, the difference between the altruism we find on a day-to-day basis among many people, and the altruism proposed by Kabbalists is the *intention*. In "Peace in the World," Baal HaSulam mentions the present type of altruism: "I am not saying that the singularity in us [the sensation that each of us is unique] will never act in us in a form of bestowal. You cannot deny that among us are people whose singularity operates in them in the form of bestowal upon others, too, such as those who spend all their money for the common good, and those who dedicate all their efforts to the common good."[147]

But when such people do good to others, they do it because it makes them feel good. In that, they are the same as any self-centered person—egoists who like to give. If they were born with a nature of enjoying hurting others, they would hurt others just as readily as they now give.

Baal HaSulam, and especially his son, Baruch (Rabash) suggest a completely different motivation for doing good to others. They suggest that people who agree on the purpose of Creation and wish to achieve it come together and do good for each other *in order to become like the Creator*. Clearly, they are just as egoistic as the rest of us, but their *goal* is different.

In "practicing" altruism in order to become it, these people discover their true nature, the nature of the Creator, and thus acquire the ability to freely choose between them. And just as children learn by imitation, gradually improving as they "practice," people who wish to be like the Creator "practice" being givers until they acquire that nature, and thus achieve the purpose of Creation.

THE DEBUT OF THE ZOHAR

Several times in this book we mentioned Item 38 in Ashlag's "Introduction to The Book of Zohar," stating that "Man [Stage Four], who can feel others, becomes needy of everything that others have, ...and is thus filled with envy to acquire everything that others have. When he has a hundred, he wants two hundred, thus his needs forever multiply until he wishes to devour all that there is in the whole world."[148]

But earlier in the introduction (Item 25), Ashlag writes, "Since the Thought [of Creation—the Creator's goal] was to delight His creatures, He *had* to create an overwhelmingly exaggerated desire to receive all that bounty, which is in the Thought of Creation [to give us unbounded pleasure]."[149] And he continues, "If the exaggerated will to receive perished from the world, the Thought of Creation would not be realized—meaning the reception of all the great pleasures that He thought to bestow upon His creatures—for the great will to receive and the great pleasure go hand in hand. And to the extent that the desire to receive it diminishes, so diminish the delight and pleasure from receiving."[150]

Hence, if we want to become Creator-like, we must not diminish our desires. But if we do not diminish our desires, then our ability to eliminate self-centeredness and become Creator-like will fail if all we have in our medicine cabinet are the old remedies of religious fanaticism, oppression, tyranny or any other of the old means of discipline. Those methods were good for "taming" the desire to receive in its earlier stages, but they will not suffice for today's level of desire to receive.

A new method, a fresh code of action is required, something that will not try to suppress the insuppressible, but will harness the new powers that extreme egoism evokes to improve life, instead of destroying both humankind and our pathogenic egocentricity.

In Stage Three of the evolution of desires, our envy has created an interconnected and interdependent world where we compete against, yet depend on each other for sustenance. In the previous chapter, we quoted Ashlag, who wrote, "Because each person in the world draws his life's marrow and his livelihood from all the people in the world, he is coerced to serve and to care for the well-being of the whole world."

We also quoted McGrew's statement: "This [single global system] defines a far more complex condition, one in which patterns of human interaction, interconnectedness, and awareness are reconstituting the world as a single social space." These quotes accurately reflect our situation at the start of the 21st century: we are tied together, and hateful of each other.

This state of simultaneous interdependence and competitiveness has brought us to a situation in which we are neither willing to negotiate with each other—as Sahtouris explains we must—nor capable of separating from each other, as did Abraham when he departed Babel. Yet, despite our self-centeredness, our interdependence *dictates* that we somehow find a way to collaborate. Thus, it seems that the only way out of this deadlock is to—as Ashlag put it—learn how to "Serve and to care for the well-being of the whole world."[151]

As we said before, the recent rise of narcissism is not a coincidence, but the result of the emergence of Stage Four in the evolution of desires. In Kabbalah, this stage is also called "the last generation." The term, "last generation," does not mean that this generation will see all of humanity become extinct. On the contrary, in the last generation, humanity should begin to *truly live* by discovering its actual vocation—becoming Creator-like. The term, "last generation," means that it is to be the last generation before the beginning of the general correction, when all of humanity discovers life's driving force—the Creator. *The Book of Zohar*, as we said in Chapter 8, describes this generation in the following way: "At the end of days, in the last generation, when your composition [*The Zohar*] appears below [in our world], because of it, you will set the land free [liberate the desire from egoism, correct it]."[152]

Descriptions of the events that will unfold in the last generation abound, the majority of them predicting humanity's doom, offering a plethora of explanations as to why we are to become extinct. Back in 1992, Chick Publications published a cartoon gospel titled, "The Last Generation." I believe that the spirit of the cartoon is best reflected in the words of one of its characters: "We may be moving to our mansions in heaven soon."[153]

Another website offers "Ten Signs of the End Times." Its author states, "I believe we are the last generation."[154] The title of the book, *The Last Generation: How Nature Will Take Her Revenge for Climate Change*,[155] by science author and journalist, Fred Pearce, speaks for itself.

Kabbalists, too, designate these times as "the last generation." In fact, they refer to the end of the 20th century as the end of the last generation, and imply that henceforth enters the state of the Age of Correction. Hillel Shklover, disciple of the great 18th century Kabbalist, The Vilna Gaon (GRA), wrote in the preface to his book, *Kol haTor* (*Voice of the Turtledove*), "From the year 1241 through the year 1990 is the period of the beginning of redemption..."[156]

Also, in Chapter 1, footnote no. 53, Shklover, writes, "[regarding] the last generation, the author [Vilna Gaon] explains that the last generation in the verse, 'That you may tell it to the last generation' (Psalms, 48:14), refers to the period beginning in 1740 through the year 1990"[157]

Likewise, in 1945, my teacher, Baruch Ashlag, told me that his father, Baal HaSulam, had commented that in fifty years, meaning in 1995, the wisdom of Kabbalah would begin its debut and people would want to study it because it would be time for it to become known. In Kabbalah, periods such as years and days often depict the passage of stages of correction, rather than the passage of physical time. Hence, when Ashlag's students asked if he was referring to physical years or to stages of correction, he replied that he was referring to physical years.

Indeed, unlike most of today's doomsday prophesies, Kabbalah predicts an entirely different scenario. Since the late Middle Ages, Kabbalists have been predicting that by using *The Book of Zohar*, people will study the law of bestowal and thus humanity will rise from despair to bliss. "Now the time dictates acquiring much possessions in the inner Torah [Kabbalah]. *The Book of Zohar* breaks new paths, sets lanes, makes a highway in the desert, ...and all its crops are ready to open the doors of redemption," wrote The Rav Kook in *Orot* (*Lights*, 1921).[158] Also, "If my people heeded me in the time of the Messiah, when evil [egoism] and heresy [oblivion to the Creator—bestowal] increase, they would delve in the study of *The Book of Zohar* and the *Tikkunim* ["Corrections," part of *The Zohar*] and the writings of the Ari all their days," wrote Rav Yitzhak Yehudah Sarfin of Komarno (1804-1976) in *Notzer Hesed* (*Keeping Mercy*).[159]

But above all else was Ashlag's momentous achievement of composing a complete translation of *The Zohar* (from Aramaic, the original language of the book, into Hebrew), a commentary of the entire book, and no fewer than four introductions to make it accessible to our generation.

THE NEED TO KNOW THE SYSTEM

We already showed how Stage Four entails a fundamentally different desire compared to the preceding stages. This stage encourages us to not only enjoy life, but to become omniscient and omnipotent like the Creator. We also explained why a new *modus operandi* is required to correct it. And by "correction," Kabbalists do not imply prohibition, oppression, or suppression of any trait, attribute, or quality of any individual. That would be repression, which would only burst twice as forcefully at the first opportunity.

As mentioned earlier, in Stage Four, the correction must be voluntary. By now, we have become so remote from Nature, so detached from the sense of life's integrity, that we simply act for ourselves, think for ourselves, and want only for ourselves. And worst of all, we do not know that another way is even conceivable. We have no awareness that living this way cannot yield a wholesome life. If this were not so, the lyrics in Little Jackie's song (which we mentioned in the Introduction), "Yes, siree, the whole world should revolve around me," would never have resonated in anyone's heart and she would never hIn every realm of life, each of us, and indeed the entire global society are striving to obtain the utmost possible, regardless of the consequences. In our personal lives, many of us succumb to what Prof. Christopher Lasch refers to as "a culture of narcissism"[160]: we promote ourselves on Facebook and on MySpace, we divorce our spouses more readily than ever, and we seek increasingly original ways to express ourselves.

In response, companies and service providers devise ever more "narcissistic" measures to cater to our self-centeredness, exponentially increasing our desires for uniqueness. Starbucks, for instance, offers close to 20,000 different combinations of coffee on their menu. Capital One has a 'Card Lab' where you can customize your credit card to include any picture you want in the background. And Facebook is so narcissism-oriented that it is built to make self promotion a breeze. Laura Buffardi and Prof. W. Keith Campbell published a research in the University of Georgia, explaining that "Narcissists are using Facebook the same way they use their other relationships—for self promotion with an emphasis on quantity of over quality"[161] (gaining as many friends as possible on their lists, despite the fact that few of these friendships yield real and lasting relationships).

And because we are so narcissistic and detached from Nature, we feel that we are not subordinate to its rules and can do whatever we want (albeit natural disasters of recent years have begun to change that view). As a result, the only way we will ever come to know how Nature works is if we make a conscious, *voluntary* effort to study it. The knowledge of how to operate within the system and yield self-interest before the interest of the system, thus receiving the support, and even cognizance of the system, is hidden behind of a veil of self-centeredness, and removing that veil is precisely the role of Kabbalah.

Kabbalists such as Rabash explain how we can imitate the Creator, imitate *true altruism,* and thus usher ourselves into the currently imperceptible level, and make it as tangible as the Nature we see before us. Without seeing and knowing the other half of reality, we will keep erring until we inflict such pain on ourselves that we will be forced to study it.

To understand how crucial this information is to our lives, consider the following scenario: You are a caveman standing in front of a white, dry, hard-as-a-rock wall inside a home. Out of the wall protrudes what appears to be a shiny gray branch made of solid rock, its trunk nowhere in sight. Then, as you stand there and gaze in bewilderment at the odd exhibit, a woman casually approaches the branch, twists it with her hand as though it were a fresh twig, and, behold! out gushes abundant water! You would probably think, "She must be a God!"

But if you could speak her language and ask her how she did it, she would explain that the "branch" is something called a "faucet," which connects to a tube that is in turn connected to a bigger tube that connects all the tubes from all the neighboring homes and continues all the way to the river. At the river, there is a big machine that pumps out the water and sends it through the tubes to all the homes in the neighborhood.

Without understanding the entire system, we are like that caveman gazing at the visible world in bewilderment, trying to figure out how it all works. And without learning from those who already know, the Kabbalists, we have about as much of a chance to figure out the system as that caveman had of figuring out that the water traveled in hoses from the river to the homes.

However, and this is important, all the above does not mean that we must *all* study Kabbalah or *The Zohar*. It only means that we will have to know the basic laws of life in our world today, which is a connected world. Similarly, we do not need a PhD in physics to know that we cannot stay in midair because there is force that pulls everything to the ground, which makes jumping off high places extremely dangerous.

However, just as it is good to know where we can learn more about the law that pulls us down because there might be useful things that we can do with it, it is good to know where we can learn more about the hidden part of reality, because knowing it might hold some benefits for us. Therefore, learning how to operate in an interconnected and interdependent world will be the topic of our next and final chapter.

11
A New Modus Operandi

Thus far, we have glanced very broadly at the history and structure of the world as explained by Kabbalah. We have described Kabbalah's view of reality as a single entity, with humans representing the highest level of existence, in the sense that we possess the most intense and most narcissistic desire to receive. It is now time to outline what humanity can do to shift the negative trend, considering that we are irreversibly interdependent and interconnected. And while it is beyond the scope of this book to outline a detailed "bailout" plan for humanity's present and future crises, we would like to point out some solutions that we believe could be implemented on a broad scale, and if done right, resolve most of our problems.

Although humanity has little experience operating as a global system, since we are used to defining ourselves as individuals or members of factions of society, from family to nation-state, the current situation necessitates that we expand our view. Most of the political and financial leaders in the world already acknowledge this requirement.

Kofi Annan, former Secretary General of the United Nations, for example, addressed this issue in a message to the First Annual Interdependence day on September 12, 2004: "A new era is upon us. In the future...the world will be transformed...by the forces of globalization and the growing interdependence of the world's peoples. ...the more interdependent we become, the more decisions have to be taken not by one nation state alone, but by many, acting together. Unless it is properly managed, this process can entail a 'democratic deficit,' as decision makers are further removed from and less accountable to the people whose lives are affected. So the challenge for all of us is to manage our interdependence in ways that bring people in, rather than shutting them out. Citizens need to think and act globally, so as to influence global decisions"[162] (*The Interdependence Handbook: Looking Back, Living the Present, Choosing the Future*, edited by Sondra Myers and Benjamin R. Barber).

More recently, in September, 2008, British Prime Minister Gordon Brown addressed the issue of globalization and global responsibility in several statements. "Each generation believes it is living through changes their parents could never have imagined—but the collapse of banks, the credit crunch, the trebling of oil prices, the speed of technology, and the rise of Asia—nobody now can be in any doubt that we are in a different world and it's now a global age."[163]

Later, Brown reflected on the background of globalization: "And we know that the challenges we face in this new global age didn't begin in the last week, or in the last months, but in fact reflect deeper changes in our world."[164]

Brown is right in saying that a deeper change is underway. When Stage Four sets in, it induces collectivism and globality (note that *globality* is a condition in which the process of globalization is complete, unlike *globalism*, which compares with imperialism and nationalism). The last stage in human development—becoming like the Creator—cannot be achieved alone. It entails uniting all fragments of Adam's soul, and through that union constructing the quality of bestowal, which is the Creator. All of us are parts of the desire that was created in Stage Four, the desire intended to achieve the purpose of creation—being Creator-like. Hence, we must reconstruct that broken desire, broken soul, together. And to do that, we must reunite in the sense that we become fully aware of our oneness and begin to experience the truth of our interconnectedness in a very real way, instead of our current, limited perception of it.

COLLABORATION AND SELF-FULFILLMENT

To experience interconnectedness, we must act on it, and not just at the global or national level. Awareness of our interdependence should be a constant in our decision-making processes and an inseparable part of the resulting actions. We must learn how to think as a unit that consists of many collaborating individuals, rather than as disparate, separate, and randomly interacting individuals. And to do that, we must begin to recognize the benefits of collaboration.

Numerous experiments have already been made on the benefits of collaboration in the education system. In an essay called, "An Educational Psychology Success Story: Social Interdependence Theory and Cooperative Learning," University of Minnesota professors, David W. Johnson and Roger T. Johnson present a compelling case for the "social interdependence" theory. In their words, "More than 1,200 research studies have been conducted in the past 11 decades on cooperative, competitive, and individualistic efforts. Findings from these studies have validated, modified, refined, and extended the theory."[165]

Johnson and Johnson explain that "Social interdependence exists when the outcomes of individuals are affected by their own and others' actions."[166] This is contrary to dependence, in which party A depends on party B, but party B may not depend on party A. "There are two types of social interdependence," they assert, "positive (when the actions of individuals promote the achievement of joint goals) and negative (when the actions of individuals obstruct the achievement of each other's goals)."[167]

If we reflect on the Christakis-Fowler experiment mentioned in the previous chapter, and consider the long standing assertion of Kabbalists that we are all factions of a single entity, it becomes clear that today, acting individualistically is not only unwise, but is nothing short of a time bomb. Such attitudes do not recognize the reality of total globalization among all members of the human race. And when we ignore this fact, reality harshly sets us straight, as the 2008 financial crisis so clearly demonstrated.

After establishing the meaning of interdependence, Johnson and Johnson went on to compare the effectiveness of cooperative learning to the commonly used individual, competitive learning. The results were unequivocal. In terms of individual accountability and personal responsibility, they concluded, "The positive interdependence that binds group members together is posited to result in feelings of responsibility for (a) completing one's share of the work and (b) facilitating the work of other group members. Furthermore, when a person's performance affects the outcomes of collaborators, the person feels responsible for the collaborators' welfare as well as for his or her own. Failing oneself is bad, but failing others as well as oneself is worse."[168] In other words, positive interdependence turns individualistic people into caring and collaborating ones, the complete opposite of the current trend of growing individualism to the point of narcissism.

Johnson and Johnson define positive interdependence as "A positive correlation among individuals' goal attainments; individuals perceive that they can attain their goals if and only if the other individuals with whom they are cooperatively linked attain their goals."[169] They define negative interdependence as "A negative correlation among individuals' goal achievements; individuals perceive that they can obtain their goals if and only if the other individuals with whom they are competitively linked fail to obtain their goals."[170] Globalization entails positive interdependence. In other words, either we all attain our goals, or none of us will.

In order to demonstrate the benefits of collaboration, the researchers measured the achievements of students who collaborated compared to those who competed. "The average person cooperating was found to achieve at about two thirds of a standard deviation above the average person performing within a competitive or individualistic situation."[171]

To understand the meaning of such deviation above the average, consider that if, for example, a child is a D-average student, by cooperating, his or her grades will leap to an astonishing A+ average. Also, they wrote, "Cooperation, when compared with competitive and individualistic efforts, tends to promote greater long-term retention, higher intrinsic motivation, and expectations for success, more creative thinking... and more positive attitudes toward the task and school."[172]

In the previous chapter, we said that by performing acts of altruism we imitate the Creator—the life-giving force that creates and propels everything that happens. We also said that just as children become grownups by imitating grownups, we will become Creator-like by imitating the Creator. Without realizing it, by collaborating, these students were imitating the law of bestowal, where self-interest yielded before the interest of their environments, which were their groups. And instead of achieving in line with the group's average capabilities, the students became straight A+ students.

Yet, what they achieved pales in comparison with the benefits that these students could achieve if they acted as they did *in order to imitate the law of bestowal*. In that case, they would discover that law and would achieve the purpose of Creation. Namely, they would become Creator-like.

Children imitate a role model that they want to become, be it a pop star, a successful athlete, or any person that they admire. Similarly, we must know that we want to be like the Creator in order to become that. It cannot happen "by mistake." As beneficial and collaborative as the behavior of the children was in the experiment, to have a lasting effect they must do it with the goal of discovering the law of yielding self-interest before the system, and aspire to live it, to become it. Otherwise, their egos will take over as all the other people are overcome by their egos, and they will lose the great benefits that social interdependence offers.

ONE OF TWO PATHS

The above section described the personal benefits of collaboration. Johnson and Johnson proved that it is more rewarding to work in a group than alone. Hence, why do we not cooperate all the time? If we are made of a desire to receive and can receive more by collaborating, then why are we not collaborating? What is it about our nature that, despite the 1,200 studies that prove it is better to work together than alone, we have not thoroughly installed these methods in our education system? And why do schools (and the entire education system), media, sports, and politics still promote competitive and individualistic behavior, extolling successful individuals? Why not extol people who promote bonding and mutuality, if evidence proves that it would work to everyone's benefit?

The reason why this is so is because in Stage Four we are no longer satisfied with achieving more. Achieving more was what we wanted in Stage Three. In Stage Four, our primary desire is to achieve *more than others*. We want to be unique and superior, just like the Creator. Thus, we may provide hard, indisputable evidence that it is better to work together than alone, but without *feeling* that this is so, our egos will not succumb to the idea. In Stage Four, solutions must first satiate the ego before we can approach daily life tactics to improve our achievements.

In regard to the above paragraph, in "Peace in the World," Baal HaSulam elaborates on our sense of uniqueness: "The nature of each and every person is to exploit the lives of all other people in the world for his own benefit. And all that he gives to another is only out of necessity; and even then there is exploitation of others in it, but it is done cunningly, so that his neighbor will not notice it and concede willingly. The reason for it," he explains, "is that... because man's soul extends from the Creator, who is One and Unique [referring to the single law of bestowal that creates and sustains the world]... man... feels that all the people in the world should be under his own governance and for his own private use. And this is an unbreakable law. The only difference is in people's choices: One chooses to exploit people to satisfy lower desires, and one by obtaining government, while the third by obtaining respect. Furthermore, if one could do it without much effort, he would agree to exploit the world with all three combined—wealth, government, and respect. However, he is forced to choose according to his possibilities and capabilities. This law can be called, 'the law of singularity in man's heart.' No person escapes it, and each and every one takes his share in that law."[173]

On October 15, 2006, Sam Roberts of *The New York Times* published a story titled, "To Be Married Means to Be Outnumbered,"[174] where he referred to a census we discussed in the Introduction. The story revealed that "Married couples, whose numbers have been declining for decades as a proportion

of American households, have finally slipped into a minority...
The American Community Survey, released... by the Census
Bureau, ...found that 49.7 percent [of] households in 2005
were made up of married couples... down from more than 52
percent five years earlier." Moreover, revealed Roberts: "The
numbers of unmarried couples are growing. Since 2000, those
identifying themselves as unmarried opposite-sex couples rose
by about 14 percent, male couples by 24 percent and female
couples by 12 percent."

Baal HaSulam wrote quite extensively of the kind of life he
envisioned for people during the correction period, which began
in the latter part of the 20th century, as noted in the previous
chapter. In his "Writings of the Last Generation," Baal HaSulam
describes two ways by which humanity can reveal the law of
bestowal (which he refers to as "completeness" in this text)—the
path of light (bestowal) and the path of suffering. In his words,
"I have already said that there are two ways to discover the
completeness: the path of light and the path of suffering. Hence,
the Creator... gave humanity technology [which he does not say
is a bad thing in and of itself], until they invented the atom and
the hydrogen bombs. If the total ruin that they are destined to
bring is still not evident to the world, they [people] can wait for
a third world war, or a fourth one and so on. The bombs will
do their thing and the relics after the ruin will have no other
choice but to take upon themselves this work, where individuals
and nations do not work for themselves more than is necessary
for their sustenance, while doing everything else for the good of
others [according to the law of yielding self-interest before the
interest of the host system]. If all the nations of the world agree
to that, there will no longer be wars in the world; no person will
be concerned with his own good whatsoever, but only with the
good of others."[175]

Ashlag concludes this section with the words, "If you take the path of light, all will be well, and if you do not, then you will tread the path of suffering. In other words, wars will breakout with atom and hydrogen bombs, and the entire world will seek counsel to escape the war. Then they shall come to the Messiah [the force that pulls out of egoism, as explained in Chapter 8]... and he will teach them this law [of bestowal]."[176]

In summary, we *have* to become Creator-like! The only question is, "How do we wish to go about it?"

TAKING THE LAW OF NATURE AS A GUIDE

Assuming that a world war, much less an atomic one, is hardly a desirable option to consider, it is far better to explore the other possibility—the path of light. In Chapter 2, we said that the term, "light," refers to the sensation of ample delight that the desire to receive (which is our essence) experiences when it is filled with pleasure. Now we can add that the pleasure is sensed when we achieve the quality of the Creator, because this is what we want at our current stage of To obtain the light, we need not all study Kabbalah. We need to *imitate* the law of bestowal, knowing what it is that we are imitating and what we want to achieve by doing so. Just as children learn by imitating grownups, to acquire the quality of bestowal we need to imitate it in our relationships.

The book, *Shamati* (*I Heard*), contains talks that Yehuda Ashlag gave on various occasions. These were put on paper by his son, and great Kabbalist in his own right, Rabash. In talk no. 7, titled, "Habit Becomes Second Nature," Baal HaSulam states, "Through accustoming oneself to some thing, that thing becomes second nature for that person. ...This means that although one has no sensation of the thing [referring to the law of bestowal], by accustoming to that thing, one can still come to experience it."[177]

Imitating the quality of bestowal in order to obtain it may seem simplistic or naïve, but at our current level of self-centeredness it is quickly becoming impossible to relate positively to *any* person, except, as we quoted Ashlag, "Out of necessity; and even then there is exploitation of others in it, but it is done cunningly, so that his neighbor will not notice it and concede willingly."[178]

The solution, according to Baal HaSulam in *The Nation*, the paper Baal HaSulam published in 1940 (see Chapter 2, "Stages Zero and One"), is to "Attend school once more."[179] In other words, we need to learn about the basic laws of Nature and the goal of life:

1. At the basis of the whole of reality lies a desire to bestow, a.k.a. "the Creator."

2. At the bottom of man's heart lies a desire to receive what the desire to bestow wishes to impart—total power, total awareness, and total governance.

3. To receive the endowments described in the above item, one must become like the desire to bestow— the Creator—itself and thus automatically have what the Creator has. This is the goal of life.

Once we realize that there is a greater reward in a cooperative conduct than in individualism it will be easy to collaborate and to share, as described in Johnson and Johnson's essay. Without this awareness, our egos will make it increasingly hard to do so, and will eventually prevent any such possibility, despite the obvious benefits.

In a rather picturesque depiction of the ego's control over us, Baal HaSulam wrote that egoism is the evil inclination, and that it approaches us holding a sword with a poisoned tip. He describes how one is enchanted by the sword and becomes a slave to one's ego, while knowing that "...in the end, the bitter drop at the tip of the sword reaches him, and this completes the separation [from the Creator] to the last spark of his breath of life"[180]

The above paragraph might induce the conclusion that Baal HaSulam believes all is lost and we are doomed to suffer. But this is not the case. We have a highly effective means that we can use to our advantage: society. We already mentioned the influence of society in brief, but in truth, society has such power over us that it can mold us into anything it desires.

In *The Writings of Baal HaSulam*, Ashlag asserts, "The greatest of all imaginable pleasures is to be favored by the people. It is worthwhile to spend all of one's energy and corporeal pleasures to obtain a certain amount of that delightful thing. This is the magnet that the greatest in all the generations have been lured after, and for which they trivialize the life of the flesh."[181]

Therefore, to alter our social behavior, we must change our *social environment* from one that promotes individuality to one that promotes mutuality. Practically speaking, we can use the media to show how group work yields better results than individual work, how competition is detrimental to one's career, health, and even, ultimately, wealth. If we think that it is impossible, it's because the media is currently telling us that it's impossible. But what if it told us otherwise? We need to act out the corrected state and the corrected state will manifest itself.

Human nature is egoistic, so we naturally tend toward isolation and competition. But Nature's essence is holistic, so it is naturally inclined toward cooperation. Acting like the rest of Nature, despite our inborn tendency not to, is our free choice, and this is what will make us similar to Nature (the Creator). Using the social environment to encourage us to follow that direction is a tool we cannot allow ourselves to miss.

APPLYING
THE CHANGE TO LIFE

We need not look very far to find ways to implement the principles of bestowal to life. When Christakis and Fowler made their astounding discoveries regarding the impact of the human network, they were relying on existing data that was collected for a very different purpose: the Framingham Heart Study, a project to understand the roots of heart disease. The reason why the researchers of the Framingham Heart Study did not discover the implications of the network—that we "infect" each other psychologically almost as we do physically—is very simple: they were not looking for such implications.

Similarly, there are many ways to observe the effects of the law of bestowal, if we only look for them as we analyze existing data. The Social Interdependence Theory, displayed here by Johnson and Johnson, is one way of observing its effect on systems, but there are many other ways to observe it. In my discussions with Professor Ervin Laszlo, philosopher of science and system theorist, we were in complete agreement because every system theorist knows that no system can persist without its parts yielding to the interests of the system.

Similar agreement transpired in my conversations with evolutionary biologist, Elisabet Sahtouris, with primatologist, Jane Goodall, and with many others. In fact, any physician, network scientist, or biologist knows that to keep a system in balance, or "homeostasis," the interests of the system must override those of its parts. Each field of science refers to this principle by a different name, and Kabbalah calls it "the law of bestowal." Essentially, however, these are different names pointing to different manifestations of the same law.

Earlier in the book, we described the views of pioneer quantum physicist, Werner Heisenberg, "Unity and complementarity constitute reality,"[182] as well as similar quotes by his contemporary, Erwin Schrödinger, and by Albert Einstein. Other contemporary sciences, such as network science, and of course, the chaos theory with its (by now) cliché, the butterfly effect, take connectivity and interdependence as a given.

On the negative side, the effects of not following the law of bestowal are evident. The growing alienation in society and the escalating isolationism on the international level, as demonstrated by publications such as Christopher Lasch's, *The Culture of Narcissism*,[183] Twenge and Campbell's *The Narcissism Epidemic*,[184] and Joseph Valadez and Remi Clignet's essay, "On the Ambiguities of a Sociological Analysis of the Culture of Narcissism,"[185] clearly demonstrate our poor social health.

Indeed, the adverse effects of narcissism are beginning to show on the international level despite repeated declarations supporting unity, such as the ones quoted earlier in this chapter. On December 3, 2009, an Associated Press news item declared, "Isolationism soars among Americans."[186]

A poll by the Pew Research Center survey found that "Americans are turning away from the world, showing a tendency toward isolationism in foreign affairs that has risen to the highest level in four decades."[187] The poll also found that "49 percent told the polling organization that the United States should 'mind its own business' internationally and let other countries get along the best they can on their own."

Another, even more disturbing aspect of our alienation is hunger. Earlier in the book we shared the alarming statistics that more than one billion people worldwide are hungry. But perhaps even more surprising is the fact that *within the United States*, "The number of households characterized as having 'very low food security'—meaning food intake was reduced because of a lack of money—jumped from 4.7 million in 2007 to 6.7 million in 2008, and the number of children in this situation rose from 700,000 to almost 1.1 million," according to a *Los Angeles Times* editorial titled, "A rising tide of hunger."[188]

Just ten days prior, *The New York Time's* Jason DeParle published an essay titled, "Hunger in U.S. at a 14-Year High," stating, "The number of Americans who lived in households that lacked consistent access to adequate food soared last year, to 49 million, the highest since the government began tracking what it calls 'food insecurity'... the Department of Agriculture reported."[189]

But the problem is not lack of food; it is lack of mutual responsibility and a basic understanding that we are going to survive together or perish together because this is the law of life. There is no shortage of food in the U.S., but there is certainly a shortage of collaboration. On January 1, 2009, Andrew Martin of *The New York Times* described "a glut of milk—and its assorted byproducts, like milk powder, butter and whey proteins—that has led to a precipitous drop in prices."[190] Martin explained that the milk products were being stored in warehouses and deliberately kept out of the stores to prevent a further drop in prices. How hard could it be to find a satisfactory arrangement to guarantee the financial stability of farmers while not depriving millions of Americans of a staple such as milk? Clearly, if we followed the law of bestowal, even if only within the U.S., such absurdities would not happen.

Thus, both manifestations of following the law of bestowal (and its mundane attire of yielding self-interest before the interest of the system) and manifestations of breaking this law abound in our world. All we need in order to realize its inclusiveness is to become aware of its existence. And for that, we need to start with education.

REARING CHILDREN TOWARD MUTUAL RESPONSIBILITY

The great Kabbalist, Rav Avraham Kook (1865-1935, best known as the first Chief Rabbi of Israel), wrote on several occasions that in the 20th century we should all study Kabbalah. However, in some cases he explicitly said that we must cultivate new ways of doing so. In a letter, published in the book, *Letters*, he wrote, "I wish to awaken all the young people who wish to be encouraged toward spiritual life. We must acquire literary skills, a lively and colorful style, and prose, and allegories. If there is anyone among us who feels inclined toward poetry, let him not neglect his gift. ...We must prepare our timely weapon—the pen. We must translate the whole of our sacred treasure into a contemporary style, ...to bring it closer to our contemporaries."[191]

Similarly, in "Introduction to the book, *Panim Me'irot uMasbirot*," Baal HaSulam wrote, "We must establish seminaries and compose books to hasten the distribution of the wisdom."[192]

By making Kabbalistic texts accessible, Ashlag and Kook wished to make its wisdom popular so people would know the basic laws of life and would know how to conduct themselves against the mounting egoism. If their advice had been heeded, Kabbalah would have been made popular a hundred years ago, and people would know about the law of bestowal before the atrocities of World War II took place.

However, there is a rule in Kabbalah: Never look back in remorse—only take what you can from the past to prepare for the future. It is certainly not too late to begin to inform people of the hidden laws of Nature, which affect our lives in a very real way. Just as everyone studies the basic laws of physics and biology at school, youths today should learn the basic laws of Nature.

In school, the principles of the Social Interdependence Theory can be a wonderful beginning, if combined with the understanding of the ultimate goal of life. If children apply these laws to their schooling, they will benefit in many more respects than just education. In the previously mentioned research, Johnson and Johnson came to several far-reaching conclusions:

- "Cooperative experiences predicted cooperative predispositions, the absence of individualistic predispositions, and engagement in prosocial behavior. Cooperative predispositions predicted the engagement in prosocial behavior and the absence of engaging in harm-intended aggression."[193]

- "If schools wish to prevent bullying and increase prosocial behaviors, the use of cooperative learning and efforts to help students become more predisposed to engage in cooperation seem to be important strategies."[194]

- "Working cooperatively with peers and valuing cooperation result in greater psychological health than do competing with peers or working independently. Cooperative attitudes were highly correlated with a wide variety of indices of psychological health. More specifically, cooperativeness is positively related to emotional maturity, well-adjusted social relations, strong personal identity, ability to cope with adversity, social competencies, basic trust and optimism about people, self confidence, independence and autonomy, higher self-esteem, and increased perspective taking skills." On the other hand, "Individualistic attitudes were negatively related to a wide variety of indices of psychological health, especially a wide variety of pathology, basic self-rejection, and egocentrism."[195]

- "Social interdependence theorists note that both positive and negative interdependence create conflicts among individuals." However, "In cooperative situations, conflicts occur over how best to achieve mutual goals. In competitive situations, conflicts occur over who will win and who will lose."[196]

In their conclusions, they also include suggestions concerning the structure of what they call a "Cooperative School." While elaborating on these is beyond the scope of this book, it is important to note how broad the effectiveness of cooperative learning seems to be: "Approximately 65% of the research... conducted on cooperative learning represents field studies demonstrating its effectiveness in a wide range of classes, subject areas, grade levels, and students. The use of cooperative learning procedures by so many different teachers, in so many different subject areas and settings, in preschool through adult education, with so many varied tasks and students, and in so many different countries and cultures, validates the theory and the clarity of the conceptual definitions."[197]

However, as effective as these teaching methods might be, they can neither succeed nor even be accepted without also teaching children the law of bestowal, and that life's purpose is to ultimately resemble that law, with all the benefits included in this similarity. Without providing this information, man's perpetually growing egotism will eventually subdue any attempt at collaboration and will increasingly isolate people, as it has been doing for the past several decades.

As Ashlag described it, we will put the sword to our tongues, to taste the sweet nectar of narcissism, and die. Indeed, the title alone of Jean M. Twenge's book, *Generation Me: Why Today's Young Americans Are More Confident, Assertive, Entitled–and More*

Miserable Than Ever Before,[198] clearly expresses the ego-trap (or is it ego-trip) of our time.

In addition to the collaborative school environment, and the efforts to inform youth of life's purpose to motivate them toward change, the principles taught at school should be applied domestically. Otherwise, the clash between school values and home values will sentence all attempts to failure.

CULTIVATING A COOPERATIVE ENVIRONMENT

In the previous chapter, we mentioned Ashlag's essay, "The Freedom," where he wrote that one's thoughts are a reflection of one's environment. This is why young people's domestic environments should match the values of cooperation that schools should promote. A publication by the U.S. Department of Education, titled, "Media Guide–Helping Your Child Through Early Adolescence," stated, "It's hard to understand the world of early adolescents without considering the huge impact of the mass media on their lives. It competes with families, friends, schools, and communities in its ability to shape young teens' interests, attitudes, and values."[199] Regrettably, the majority of interests that the media shapes are antisocial.

An online publication by the University of Michigan Health System, for example, states that "Literally thousands of studies since the 1950s have asked whether there is a link between exposure to media violence and violent behavior. All but 18 have answered, 'Yes.' ...According to the AAP (American Academy of Pediatrics), 'Extensive research evidence indicates that media violence can contribute to aggressive behavior, desensitization to violence, nightmares, and fear of being harmed.'"[200]

To understand how much violence young minds absorb, consider this piece of information from the publication: "An average American child will see 200,000 violent acts and 16,000 murders on TV by age 18."[201] If this number does not seem alarming, consider that there are 6,570 days in eighteen years. This means that on average, by age eighteen a child will have been exposed to slightly more than thirty acts of violence on TV, 2.4 of which are murders, *every day of his or her young life.*

On the same note, in their book, *Development Through Life: A Psychosocial Approach*, published in 2008, Barbara M. Newman, PhD, and Philip R. Newman describe how "Exposure to many hours of televised violence increases young children's repertoire of violent behavior and increases the prevalence of angry feelings, thoughts, and actions. These children are caught up in the violent fantasy, taking part in the televised situation while they watch."[202] If we consider that children learn by imitation, we can only imagine what irreversible harm watching violence causes them.

In a capitalist country, the government does not enforce laws that prohibit violence on TV and other media. At best, the government might strive to restrict it, but the above statistics clearly indicate that these efforts are grossly ineffective. The solution should come from the people, not from government. People have to decide what they want to watch on TV, and to do that, they must decide what kind of individuals they want to be, what goals they wish to achieve, but most important, what sort of adults they want their children to become and in what kind of world they want them to grow.

When parents decide that they want their children to grow up with a hopeful future, that they do not want them to join the growing ranks of depressed youths already numbering (according to the National Mental Health Information Center) "One in every five young people at any given time,"[203] then the change will take place. TV, movies, Internet, and every other means of mass media lives and dies by its ratings. When consumers decide that they want nonviolent media, then producers, screenplay writers, and advertisers will know how to create a whole new repertoire of nonviolent films that promote cooperative behavior, as mentioned in the section "Taking the Law of Nature as a Guide."

The media is a learning aid and a democratic one in the sense that it truly depends on the viewers' favor. While it is controlled by a relatively small number of people who have their own interests regarding what to air and what not to air, at the end of the day, the media still shows us what we want to see, or else the industry would go bankrupt. Because the majority of today's people are more narcissistic than ever, so is the nature of mass media programs. And because we are becoming *increasingly* self-centered, the mass media *increasingly* caters to values of entitlement and isolationism.

Yet, isolationism and narcissism are unsustainable in an interdependent world. They are to society as cancer is to the body. The solution, therefore, is to find a way to harness our intensifying desires toward socially productive directions, which in the end are personally rewarding, as well. This is the only way we can rise above our growing egoism and unite.

The solution that Kabbalah offers is to use the newly acquired awareness of the global system that is humanity, and to teach the law bestowal that sustains it and (most important) the goal of life. The reward will be, as said above, "total power, total awareness, and total governance" (of ourselves, of our lives, and of the world). But this will happen only if and when we *choose* to unite. In doing so, we will achieve the goal of existence—the Thought of Creation—and united, we will become like our Creator.

And we can choose to do so after many painful "persuasions" on the part of Nature, or after self-persuasion using the environment, the principle of imitation, and the awareness of our social interdependence. In the essay, "The Peace," Baal

HaSulam describes two kinds of people—those who advance toward life's purpose willingly and knowingly, and reap the benefits, and those who advance unwillingly, unknowingly, and reap agony. In his words, "There is a great difference and a great distance between them, meaning 'knowingly and unknowingly.' The first type ... stormy waves come upon them, through the strong wind of development, and push them from behind, forcing them to step forward. Thus, their debt [development toward attainment of the goal knowingly] is collected against their will and with great pains ... which push them from behind. But the second type pay their debt ... of their own accord, by repeating the actions that hasten the development [imitation, influence of the environment]. ... They chase it of their own free will, with the spirit of love. Needless to say, they are free from any kind of sorrow and suffering like the first type," as well as "Hasten the [attainment of] desired goal."[204]

BIBLIOGRAPHY

Amani El-Alayli and Messe Lawrence A., "Reactions Toward an Unexpected or Counternormative Favor-Giver: does it matter if we think we can reciprocate?" *Journal of Experimental Social Psychology* 40.5 (September 2004)

Anthony McGrew, "A Global Society?" in *Modernity and Its Futures*, ed. Stuart Hall. UK: Polity Press in association with Blackwell Publishing Ltd and The Open University, 1992.

Arias Elizabeth, Ph.D., Division of Vital Statistics, "United States Life Tables, 2004," *National Vital Statistics Report* (NVSS) 56 no. 9 (December 28, 2007).

Ashlag Baruch Shalom (Rabash), *The Writings of Rabash*. Israel: Ashlag Research Institute, 2008.

Ashlag Yehuda, *Kitvey Baal HaSulam* (*The Writings of Baal HaSulam*). Israel: Ashlag Research Institute, 2009.

Ashlag Yehuda, *Shamati* (*I Heard*), trans. Chaim Ratz. Canada: Laitman Kabbalah Publishers, 2009.

Ashlag Yehuda, *Talmud Eser Sefirot* (*The Study of the Ten Sefirot*). Israel: Ashlag Research Institute, 2007.

Avraham Ben Mordechai Azulai, *Ohr HaChama* (*Light of the Sun*).

Burg Bob and David John, The Go-Giver: A Little Story About A Powerful Business Idea. USA: Portfolio, 2007.

Babylonian Talmud, Masechet [Tractate] Yoma, Masechet Hagigah, Masechet Yevamot, Masechet Kidushin.

Buffardi Laura E., Campbell W. Keith, "Narcissism and Social Networking Web Sites," *Personality & Social Psychology Bulletin* 34 (July 3, 2008): 1303-1314. doi:10.1177/0146167208320061.

Calaprice Alice, *The New Quotable Einstein*. USA: Princeton University Press, 2005.

Chaim ibn Attar, *Ohr HaChaim* [*Light of Life*], *Bamidbar* [Numbers], Chapter 23.

Collett Jessica L. and Morrissey Christopher A., "The Social Psychology of Generosity: the state of current interdisciplinary research." USA: University of Notre Dame, October 2007.

Cordovero Moshe (RAMAK), Know the God of Thy Father.

Darwin Charles, *The Works of Charles Darwin, Volume 16: The Origin of Species*, 1876. NY: NYU Press; Volume 16 edition, February 15, 2010.

Dawkins Richard, *The Selfish Gene*. New York: Oxford University Press Inc., 1989.

Della Mirandola Giovanni Pico, *De Hominis Dignitate Oratio* (*Oration on the Dignity of Man*). Italy: Feltrinelli, 2000.

Diamond Jared, Guns, Germs, and Steel: The Fates of Human Societies. NY: Norton & Company, 1997.

Elimelech of Lizhensk, *Noam Elimelech* (*The Pleasantness of Elimelech*), *Likutei Shoshana* ("Collections of the Rose") (First published in Levov, Ukraine, 1788).

Estelami Hooman and De Maeyer Peter, "Customer reactions to service provider overgenerosity," *Journal of Service Research* 4, no. 3 (Feb 2002): 205-216

Frankl Viktor E., *Man's Search for Meaning*, trans. Ilse Lasch. Boston: Beacon Press, 2006.

Hillel Shklover, Kol haTor (Voice of the Turtledove).

Horgan John, The End of Science: Facing the Limits of Knowledge in the Twilight of the Scientific Age. New York: Broadway Books, 1997.

Hurley S. and Chater N. (Eds.), *Perspectives on Imitation: From Neuroscience to Social Science*. Cambridge, MA: MIT Press, 2005.

Isaac Luria (ARI), *Tree of Life*.

Johnson David W. and Johnson Roger T., "An Educational Psychology Success Story: Social Interdependence Theory and Cooperative Learning," *Educational Researcher* 38 (2009): 365-380, doi: 10.3102/0013189X09339057

_____, *Kabbalah for the Student*, ed. Gilad Shadmon, trans. Chaim Ratz. Canada: Laitman Kabbalah Publishers, 2009.

Lasch Christopher, The Culture of Narcissism: American Life in an Age of Diminishing Expectations. USA: Norton & Company, May 17, 1991.

Laszlo Ervin, *The Chaos Point: The World at the Crossroads*. Charlottesville, VA: Hampton Roads, 2006.

Lehrs Ernst, *Man or Matter*. London: Rudolf Steiner Press; 2nd edition, June 1985.

Leibniz Gottfried Wilhelm, *Leibniz: Philosophical Writings*. UK: Dent, Rowman and Littlefield, 1991.

Lovelock James, The Revenge of Gaia: Earth's Climate Crisis & The Fate of Humanity. New York: Basic Books, 2006.

Luntschitz Shlomo Ephraim, *Keli Yakar* [*Precious Vessel*].

Midrash Rabbah, Beresheet, Portion 38.

Midrash Rabbah, Beresheet.

Midrash Rabbah, Kohelet.

Midrash Rabbah, Shemot.

Mishnah, Masechet Hagigah.

Nanda Anshen Ruth, *Biography of an Idea*. USA: Moyer Bell, 1987.

Newman Barbara M. and Newman Philip R., *Development Through Life: A Psychosocial Approach*. Belmont, CA: Wadsworth Cengage Learning, 2008.

Pearce Fred, The Last Generation: How Nature Will Take Her Revenge for Climate Change. USA: Key Porter Books, February 2007.

Postman Neil, The End of Education: Redefining the Value of School. New York: Knopf, 1995.

Rav Avraham Kook (The Raaiah), *Igrot* (*Letters*).

Rav Moshe Ben Maimon (Maimonides), *Mishneh Torah* (*Yad HaChazakah* (*The Mighty Hand*)), Part 1, "The Book of Science."

Rav Moshe Ben Maimon (Maimonides), *Mishneh Torah* (*Yad HaChazakah* (*The Mighty Hand*), Part 1, "The Book of Science," Chapter 1.

Rav Shabtai Ben Yaakov Yitzhak Lifshitz, *Segulot Israel* (The Virtue of Israel).

Rav Yitzhak Yehudah Yehiel of Komarno, *Notzer Hesed* (*Keeping Mercy*).

Reuchlin Johannes, *De Arte Cabbalistica* (*On the Art of Kabbalah*). Hagenau, Germany: Tomas Anshelm, March, 1517.

Russell Bertrand, *History of western Philosophy*. London: Routledge Classics, 2004.

Shachmurove Yochanan and Uriel Spiegel, "A Monopoly Reason Why Autarky might Be Best for a Large Country," *The Manchester School* 73, no. 3 (2005): 269-280.

Spock Benjamin, *Baby and Child Care*. USA: Pocket Books, 2004.

Rabbi Shimon bar Yochai (Rashbi), The Book of Zohar, Tikkuney Zohar (Corrections of The Zohar).

"The Global Financial System," *eJournal USA* 14, no. 5 (May, 2009)

The Interdependence Handbook: Looking Back, Living the Present, Choosing the Future, eds. Sondra Myers and Benjamin R. Barber. NY: The International Debate Education Association, 2004.

The Rav Raiah Kook, *Orot* (*Lights*).

Twenge Jean M. and Campbell W. Keith, *The Narcissism Epidemic: Living in the Age of Entitlement*. New York: Free Press, A Division of Simon & Schuster, Inc. 2009.

Twenge Jean M., Generation Me: Why Today's Young Americans Are More Confident, Assertive, Entitled–and More Miserable Than Ever Before. USA: Free Press, March 6, 2007.

Tyler Miller G. and Spoolman Scott, *Living in the Environment: Principles, Connections, and Solutions*. Belmont, CA: Books/Cole, Cengage Learning, 2008.

Valadez Joseph and Clignet Remi, "On the Ambiguities of a Sociological Analysis of the Culture of Narcissism" *Sociological Quarterly*, vol. 28, 4 (Dec. 1987): 455–472.

Von Goethe Johann Wolfgang, *Materialien zur Geschichte der Farbenlehre*. Stuttgart, Germany: Gotta'sche Buchhandlung, 1833.

Whiston William, *The Works of Flavius Josephus*. UK: Armstrong and Plaskitt AND Plaskitt & Co., 1835.

Wilber Ken, Quantum Questions: Mystical Writings of the World's Great Physicists. USA: Shambhala Publications, Inc., 1984.

Zalman Elijah ben Shlomo (The Vilna Gaon (GRA)), *Even Shlemah* (*A Perfect and Just Weight*). Israel: *Yofi* (Beauty) Publishing, 2007.

NOTES

1   Global Health Council, "Global View" (2009), http://www.globalhealth.org/
infectious_diseases/global_view/

2   James Lovelock, *The Revenge of Gaia: Earth's Climate Crisis & The Fate of Humanity* (New York: Basic Books, 2006)

3   Ervin Laszlo, *The Chaos Point: The World at the Crossroads* (Charlottesville, VA: Hampton Roads, 2006)

4   Stephan Faris, "Top 10 Places Already Affected by Climate Change," *Scientific American* 54 (December 23, 2008), http://www.scientificamerican.com/article.
cfm?id=top-10-places-already-affected-by-climate-change

5   Peter Popham, "Melting snow prompts border change between Switzerland and Italy," *The Independent* (24 March, 2009), http://www.independent.co.uk/news/
world/europe/melting-snows-prompt-border-change-between-switzerland-and-italy-1653181.html)

6   World Food Programme, "Hunger Stats" (2009), http://www.wfp.org/hunger/stats

7   Sam Roberts, "To Be Married Means to Be Outnumbered," *The New York Times* (October 15, 2006), http://www.nytimes.com/2006/10/15/us/15census.
html?scp=1&sq=To%20Be%20Married%20Means%20to%20Be%20
Outnumbered&st=cse

8   Indrajit Basu, "'Native English' is losing its power," *Asia Times* (September 15, 2006), http://www.atimes.com/atimes/South_Asia/HI15Df01.html

9   Associated Press, "Recession will likely be longest in postwar era," MSNBC (March, 2009), http://www.msnbc.msn.com/id/29582828/wid/1/page/2/

10   Fareed Zakaria, "Get Out the Wallets: The world needs Americans to spend, *Newsweek* (August 1, 2009), http://www.newsweek.com/2009/07/31/get-out-the-wallets.html

11   Jean M. Twenge and W. Keith Campbell, *The Narcissism Epidemic: Living in the Age of Entitlement* (New York: Free Press, A Division of Simon & Schuster, Inc. 2009), 78

12   Jean M. Twenge and W. Keith Campbell, *The Narcissism Epidemic*, 1

13   Jean M. Twenge and W. Keith Campbell, *The Narcissism Epidemic*, 1-2

14   "World Leaders Seek Unity to Fight Financial Crisis," *The Economic Times* (September 24, 2008), http://www.usatoday.com/news/world/2008-09-24-un-financial-crisis_N.htm

15   Yehuda Ashlag, "The Essence of the Wisdom of Kabbalah," in *Kabbalah for the Student*, ed. Gilad Shadmon, trans. Chaim Ratz (Canada: Laitman Kabbalah Publishers, 2009), 21

16   David L. Goodstein (Primary Contributor), "Mechanics," *Encyclopædia Britannica*, http://www.britannica.com/EBchecked/topic/371907/mechanics

17   "A Theory of Everything," "Subatomic Particle," *Encyclopædia Britannica*, http://www.britannica.com/EBchecked/topic/570533/subatomic-particle/254800/A-theory-of-everything

18   Werner Heisenberg, quoted by Ruth Nanda Anshen in *Biography of an Idea* (USA: Moyer Bell, 1987), 224

19   Ken Wilber, *Quantum Questions: Mystical Writings of the World's Great Physicists*, (USA: Shambhala Publications, Inc., 1984), 96

20   Alice Calaprice, *The New Quotable Einstein* (USA: Princeton University Press, 2005), 206

21   Laurance Johnston, "Objective Science: An Inherent Oxymoron" (April 2007), http://brentenergywork.com/OBJECTIVE_SCIENCE_ARTICLE.htm

22   Bertrand Russell, *History of western Philosophy* (London: Routledge Classics, 2004), 243

23   Gottfried Wilhelm Leibniz, *Leibniz: Philosophical Writings* (UK: Dent, Rowman and Littlefield, 1991), 37

24   Ernst Lehrs, *Man or Matter* (London: Rudolf Steiner Press; 2nd edition, June 1985), 58-9

25   Rav Moshe Ben Maimon (Maimonides), *Mishneh Torah* (*Yad HaChazakah* (*The Mighty Hand*)), Part 1, "The Book of Science," Chapter 1, Item 1

26   Rav Moshe Ben Maimon (Maimonides), *Mishneh Torah* (*Yad HaChazakah* (*The Mighty Hand*)), Part 1, "The Book of Science," Chapter 1, Item 3

27   Yehuda Ashlag, "The Peace," in *Kabbalah for the Student*, ed. Gilad Shadmon, trans. Chaim Ratz (Canada: Laitman Kabbalah Publishers, 2009), 265

28   Rav Moshe Ben Maimon (Maimonides), *Mishneh Torah* (*Yad HaChazakah* (*The Mighty Hand*)), Part 1, "The Book of Science," Chapter 1, Item 3

29   *Midrash Rabbah*, *Beresheet*, Portion 38, Item 13

30   *Midrash Rabbah*, *Beresheet*, Portion 38, Item 13

31   Rav Moshe Ben Maimon (Maimonides), *Mishneh Torah* (*Yad HaChazakah* (*The Mighty Hand*), Part 1, "The Book of Science," Chapter 1, Item 3

32   Rav Moshe Ben Maimon (Maimonides), *Mishneh Torah* (*Yad HaChazakah* (*The Mighty Hand*), Part 1, "The Book of Science," Chapter 1, Item 3

33   Rav Moshe Ben Maimon (Maimonides), *Mishneh Torah* (*Yad HaChazakah* (*The Mighty Hand*), Part 1, "The Book of Science," Chapter 1, Item 3

34   Elimelech of Lizhensk, *Noam Elimelech* (*The Pleasantness of Elimelech*), *Likutei Shoshana* ("Collections of the Rose") (First published in Levov, Ukraine, 1788), obtained from http://www.daat.ac.il/daat/vl/tohen.asp?id=173

35   Shlomo Ephraim Luntschitz, *Keli Yakar* [*Precious Vessel*]

36   Chaim ibn Attar, *Ohr HaChaim* [*Light of Life*], *Bamidbar* [Numbers], Chapter 23, Item 8, https://sites.google.com/site/magartoratemet/tanach/orhahaym

37   Baruch Shalom Ashlag (Rabash), *The Writings of Rabash*, Vol. 1, Article no. 9, 1988-89 (Israel: Ashlag Research Institute, 2008), 50, 82, 163

38   Neil Postman, *The End of Education: Redefining the Value of School* (USA: Vintage, First Edition, 1996), 170

39   Rav Moshe Ben Maimon (Maimonides), *Mishneh Torah* (*Yad HaChazakah* (*The Mighty Hand*), Part 1, "The Book of Science," Chapter 1, Item 3

40   Yehuda Ashlag, "The Matter of Spiritual Attainment," in *Shamati* (*I Heard*), trans. Chaim Ratz (Canada: Laitman Kabbalah Publishers, 2009), 22

41   Baruch Shalom Ashlag (Rabash), *The Writings of Rabash*, Vol. 2, Article no. 9, 1988-89 (Israel: Ashlag Research Institute, 2008), 823

42   Isaac Luria (ARI), *Tree of Life*, Gate 1, Branch 2

43   Yehuda Ashlag, *Talmud Eser Sefirot* (*The Study of the Ten Sefirot*), Part 1 (Israel: Ashlag Research Institute, 2007), 19

44   Yehuda Ashlag, *Talmud Eser Sefirot* (*The Study of the Ten Sefirot*), Part 1 (Israel: Ashlag Research Institute, 2007), 31

45   Yehuda Ashlag, "*Talmud Eser Sefirot* (*The Study of the Ten Sefirot*), Part One, *Histaklut Pnimit* (Inner Reflection)," in *Kabbalah for the Student*, ed. Gilad Shadmon, trans. Chaim Ratz (Canada: Laitman Kabbalah Publishers, 2009), 729

46   Yehuda Ashlag, "Introduction to Study of the Ten Sefirot," in *Kabbalah for the Student*, ed. Gilad Shadmon, trans. Chaim Ratz (Canada: Laitman Kabbalah Publishers, 2009), 374

47   "Lightning," *Encyclopedia Britannica* (http://www.britannica.com/EBchecked/topic/340767/lightning)

48   Ashlag, "Preface to the Wisdom of Kabbalah," in *Kabbalah for the Student*, 567-572

49   Richard Dawkins, *The Selfish Gene* (New York: Oxford University Press Inc., 1989), 14

50   Ashlag, "Preface to the Wisdom of Kabbalah," in *Kabbalah for the Student*, 567-9

51   From: S. Hurley and N. Chater (Eds.), *Perspectives on Imitation: From Neuroscience to Social Science* (Vol. 2) (Cambridge, MA: MIT Press, 2005), 55-77

52   Benjamin Spock, *Baby and Child Care*, (USA: Pocket Books, 2004), 164-5

53   Ashlag, *Kitvey Baal HaSulam* (*The Writings of Baal HaSulam*) (Israel: Ashlag Research Institute, 2009), 499

54   Ashlag, "Preface to the Wisdom of Kabbalah," in *Kabbalah for the Student*, 568

55   Ashlag, "Preface to the Wisdom of Kabbalah," in *Kabbalah for the Student*, 568

56   Yehuda Ashlag, "The Giving of the Torah," in *Kabbalah for the Student*, 244

57   El-Alayli Amani and Lawrence A. Messe. "Reactions Toward an Unexpected or Counternormative Favor-Giver: does it matter if we think we can reciprocate?" *Journal of Experimental Social Psychology* 40.5 (September 2004): 633-641

58   (ibid.)

59   Ashlag, "The Giving of the Torah," in *Kabbalah for the Student*, 244

60   Ashlag, "The Essence of the Wisdom of Kabbalah," in *Kabbalah for the Student*, 25

61   Ashlag, "Introduction to the Book of Zohar," in *Kabbalah for the Student*, 128

62   (ibid.)

63   (ibid.)

64   (ibid.)

65   (ibid.)

66   Ashlag, "The Giving of the Torah," in *Kabbalah for the Student*, 244

67   Ashlag, "Preface to the Wisdom of Kabbalah," in *Kabbalah for the Student*, 571-573

68   Nobel Lecture by Ada E. Yonath, http://nobelprize.org/mediaplayer/index.php?id=1212&view=1

69   Ashlag, "Preface to the Wisdom of Kabbalah," in *Kabbalah for the Student*, 567-568

70   Yehuda Ashlag, *Talmud Eser Sefirot* (*The Study of the Ten Sefirot*), Part 1 (Israel: Ashlag Research Institute, 2007), 5

71   "Where Did All the Elements Come From??" Haystack Observatory, an interdisciplinary research center of the Massachusetts Institute of Technology (MIT) (August 11, 2005), http://www.haystack.mit.edu/edu/pcr/Astrochemistry/3%20-%20MATTER/nuclear%20synthesis.pdf

72   "Helium-3 in Milky Way Reveals Abundance of Matter in Early Universe," National Radio Astronomy Observatory (January 2, 2002), http://www.nrao.edu/pr/2002/he3/

73   Richard Dawkins, *The Selfish Gene* (New York: Oxford University Press Inc., 1989), 13

74   Lynn Margulis, Carl Sagan, Dorion Sagan (Primary Contributors), "Life," *Encyclopædia Britannica*, http://www.britannica.com/EBchecked/topic/340003/life

75   Ashlag, "Introduction to the Book of Zohar," in *Kabbalah for the Student*, 128

76   Yehuda Ashlag, *Talmud Eser Sefirot* (*The Study of the Ten Sefirot*), Parts 10-12 (Israel: Ashlag Research Institute, 2007), 865-1296

77   Ashlag, "Introduction to the Book of Zohar," in *Kabbalah for the Student*, 128

78   (ibid.)

79   (ibid.)

80   (ibid.)

81   (ibid.)

82   United States Geological Survey (USGS), "Why did the dinosaurs die out?" (May 17, 2001), http://pubs.usgs.gov/gip/dinosaurs/die.html

83   University of California Museum of Paleontology, "What Killed The Dinosaurs?" (January 2009), http://www.ucmp.berkeley.edu/diapsids/extinctheory.html

84   "Evolution Can Occur in Less Than Ten Years," *Science Daily* (June 15, 2009), http://www.sciencedaily.com/releases/2009/06/090610185526.htm

85   Wendy Zukerman, Australia's battle with the bunny, *ABC Science* (April 08, 2009), http://www.abc.net.au/science/articles/2009/04/08/2538860.htm

86   (ibid.)

87   (ibid.)

88   Louis L. Ray, "The Great Ice Age," U.S. Geological Survey (September 27, 1999), http://pubs.usgs.gov/gip/ice_age/ice_age.pdf

89   Robin Allaby, "Research pushes back history of crop development 10,000 years," University of Warwick (September 19, 2008), http://www2.warwick.ac.uk/newsandevents/pressreleases/research_pushes_back/

90   Jared Diamond, *Guns, Germs, and Steel: The Fates of Human Societies* (NY: Norton & Company, 1997)

91   Jared Diamond, "The Evolution of Religions" (Uploaded by RabidApe, May 26, 2009), http://www.youtube.com/watch?v=GWXr7pXoCTs

92   Ashlag, "Peace in the World," in *Kabbalah for the Student*, 89

93   (ibid.)

94   Douglas Adams, *Dirk Gently's Holistic Detective Agency* (NY: Pocket Books, 1987), 270

95   *Midrash Rabbah*, *Shemot* 2:4

96   RASHI commentary on exodus 19:2

97   Ashlag, *The Arvut* (The Mutual Guarantee), in *Kabbalah for the Student*, 251

98   (ibid.)

99   *Midrash Rabbah, Kohelet*, 1:13

100 Ashlag, *Kabbalah for the Student*, 54

101 Babylonian Talmud, *Masechet* [Tractate] *Yoma* p 9b

102 Babylonian Talmud, *Yoma*, p 9b

103 Babylonian Talmud, *Masechet* [Tractate] *Yevamot*, 62b

104 Johannes Reuchlin, *De Arte Cabbalistica* (Hagenau, Germany: Tomas Anshelm, March, 1517), 126

105 Giovanni Pico della Mirandola, *De Hominis Dignitate Oratio* (*Oration on the Dignity of Man*) (Italy: Feltrinelli, 2000), 148

106 William Whiston, *The Works of Flavius Josephus* (UK: Armstrong and Plaskitt AND Plaskitt & Co., 1835), 564

107 Whiston, *The Works of Flavius Josephus*, 565

108 Yehuda Ashlag, "The Love of the Creator and the Love of Man," in *Kitvey Baal HaSulam* (*The Writings of Baal HaSulam*) (Israel: Ashlag Research Institute, 2009), 486

109 Kip P. Nygren, "Emerging Technologies and Exponential Change: Implications for Army Transformation," *Parameters* (Summer 2002), 86-99, Online source: http://www.carlisle.army.mil/usawc/parameters/Articles/02summer/nygren.htm

110 G. Tyler Miller and Scott Spoolman, *Living in the Environment: Principles, Connections, and Solutions* (Belmont, CA: Books/Cole, Cengage Learning, 2008)

111 Solliberty (online name), "Did You Know? We are living in exponential times (December 9, 2008), http://www.youtube.com/watch?v=IUMf7FWGdCw

112 Ashlag, "Introduction to the Book of Zohar," in *Kabbalah for the Student*, 128

113 Elijah ben Shlomo Zalman (The Vilna Gaon (GRA)), *Even Shlemah* (*A Perfect and Just Weight*), Chapter 11, Item 3 (Israel: *Yofi* (Beauty) Publishing, 2007), 100

114 Mishnah, *Masechet Hagigah*, 2,1

115 Babylonian Talmud, *Masechet Hagigah*, p 14b

116 Babylonian Talmud, *Masechet Kidushin*, Chapter 1, p 30a)

117 *The Book of Zohar, Tikkuney Zohar* (*Corrections of the Zohar* (part of *The Zohar*), *Tikkun* (Correction) No. 6, p 24a)

118 Chaim Vital, *The Tree of Life, Introduction of Rav Chaim Vital to The Gate of Introductions*, http://www.kab.co.il/heb/content/view/frame/7948?/heb/content/view/full/7948&main

119 (ibid.)

120 (ibid.)

121 (ibid.)

122 (ibid.)

123 Avraham Ben Mordechai Azulai, *Ohr HaChama* (*Light of the Sun*), Introduction, p 81

124 Rav Moshe Cordovero (RAMAK), *Know the God of Thy Father*, 40

125 Rav Yitzhak Yehudah Yehiel of Komarno, *Notzer Hesed* (*Keeping Mercy*), Chapter 4, Teaching 4

126 Rav Shabtai Ben Yaakov Yitzhak Lifshitz, *Segulot Israel* (The Virtue of Israel), Set no. 7, Item 5

127 Johann Wolfgang von Goethe, *Materialien zur Geschichte der Farbenlehre* (Germany: Gotta'sche Buchhandlung, 1833), 83-4

128 The Wright Brothers, the Invention of the Aerial Age, "Inventing a Flying Machine," http://www.nasm.si.edu/wrightbrothers/fly/1903/triumph.cfm

129 Elizabeth Arias, Ph.D., "National Vital Statistics Reports," Vol. 56, No. 9 (December 28, 2007): 31-32

130 Richard H. Pells, Christina D. Romer (Primary Contributors), "Great Depression," http://www.britannica.com/EBchecked/topic/243118/Great-Depression

131 Charles Darwin, *The Works of Charles Darwin, Volume 16: The Origin of Species*, 1876 (NY: NYU Press; Volume 16 edition, February 15, 2010), 167

132 (ibid.)

133 *Midrash Rabah, Kohelet*, 1:13

134 Babylonian Talmud, *Masechet Sukkah*, p 52a

135 *Midrash Rabah, Kohelet*, 1:13

136 Anthony McGrew, "A Global Society?" in *Modernity and Its Futures*, ed. Stuart Hall (UK: Polity Press in association with Blackwell Publishing Ltd and The Open University, 1992), 65

137 Ashlag, "Peace in the World," in *Kabbalah for the Student*, 92

138 Ashlag, "Peace in the World," in *Kabbalah for the Student*, 93

139 (ibid.)

140 Clive Thompson, "Are Your Friends Making You Fat?", *The New York Times* (September 10, 2009), http://www.nytimes.com/2009/09/13/magazine/13contagion-t.html?_r=1&th&emc=th

141 (ibid.)

142 (ibid.)

143 (ibid.)

144 (ibid.)

145 Ashlag, "The Freedom," in *Kabbalah for the Student*, 384

146 Clive Thompson, "Are Your Friends Making You Fat?", *The New York Times* (September 10, 2009), http://www.nytimes.com/2009/09/13/magazine/13contagion-t.html?_r=1&th&emc=th

147 Ashlag, "Peace in the World," in *Kabbalah for the Student*, 96

148 Ashlag, "Introduction to the Book of Zohar," in *Kabbalah for the Student*, 128

149 Ashlag, "Introduction to the Book of Zohar," in *Kabbalah for the Student*, 122-3

150 (ibid.)

151 Ashlag, "Peace in the World," in *Kabbalah for the Student*, 92

152 *The Book of Zohar, Tikkuney Zohar* (*Corrections of the Zohar* (part of *The Zohar*), *Tikkun* (Correction) No. 6, p 24a)

153 Jack T. Chick "The Last Generation," *Chick Publications* (1992), http://www.chick.com/reading/tracts/0094/0094_01.asp

154 "Ten Signs of the End Times ," http://www.escapeallthesethings.com/last-generation.htm

155 Fred Pearce, *The Last Generation: How Nature Will Take Her Revenge for Climate Change* (USA: Key Porter Books, February 2007)

156 Hillel Shklover, *Kol haTor* (*Voice of the Turtledove*), 498

157 Hillel Shklover, *Kol haTor* (*Voice of the Turtledove*), 553

158 The Rav Raiah Kook, *Orot* (*Lights*), 57

159 Rav Yitzhak Yehuda Yehiel of Komarno, *Notzer Hesed* (*Keeping Mercy*), Chapter 4, Teaching 20

160 Christopher Lasch, *The Culture of Narcissism: American Life in an Age of Diminishing Expectations* (USA: Norton & Company, May 17, 1991)

161 Laura E. Buffardi, W. Keith Campbell, "Narcissism and Social Networking Web Sites," *Personality & Social Psychology Bulletin* 34 (July 3, 2008): 1303-1314, doi:10.1177/0146167208320061, quoted in "Facebook Profiles Can Be Used To Detect Narcissism," *Science Daily* (September 23, 2008): http://www.sciencedaily.com/releases/2008/09/080922135231.htm

162 *The Interdependence Handbook: Looking Back, Living the Present, Choosing the Future*, ed. Sondra Myers and Benjamin R. Barber (NY: The International Debate Education Association, 2004), 14

163 Gordon Brown speaks to Conference, *Labour* (September 23, 2008): http://www.labour.org.uk/gordon_brown_conference

164 (ibid.)

165 David W. Johnson and Roger T. Johnson, "An Educational Psychology Success Story: Social Interdependence Theory and Cooperative Learning," *Educational Researcher* 38 (2009): 365, doi: 10.3102/0013189X09339057

166 Johnson and Johnson, "Educational Psychology Success Story," 366

167 (ibid.)

168 Johnson and Johnson, "Educational Psychology Success Story," 368

169 (ibid.)

170 (ibid.)

171 Johnson and Johnson, "Educational Psychology Success Story," 371

172 (ibid.)

173 Ashlag, "Peace in the World," in *Kabbalah for the Student*, 89

174 Sam Roberts, "To Be Married Means to Be Outnumbered, *The New York Times* (October 15, 2006): http://www.nytimes.com/2006/10/15/us/15census.html?_r=1&scp=2&sq=more%20unmarried%20couples%20than%20married%20couples&st=cse

175 Yehuda Ashlag, *Kitvey Baal HaSulam* (*The Writings of Baal HaSulam*), "The Writings of the Last Generation," Part 1 (Israel: Ashlag Research Institute, 2009), 815

176 (ibid.)

177 Yehuda Ashlag, "What Is Habit Becomes a Second Nature in the Work," in *Shamati* (*I Heard*), trans. Chaim Ratz (Canada: Laitman Kabbalah Publishers, 2009), 38

178 Ashlag, "Peace in the World," in *Kabbalah for the Student*, 89

179 Yehuda Ashlag, *The Writings of Baal HaSulam* "The Nation," (Israel: Ashlag Research Institute, 2009), 494

180 Ashlag, "Introduction to the Book, Panim Meirot uMasbirot" (Illuminating and Enlightening Face) in *Kabbalah for the Student*, 463

181 Yehuda Ashlag, *Kitvey Baal HaSulam* (*The Writings of Baal HaSulam*), 44

182 Werner Heisenberg, quoted by Ruth Nanda Anshen in *Biography of an Idea*, 224

183 Christopher Lasch, *The Culture of Narcissism: American Life in an Age of Diminishing Expectations* (USA: Norton & Company, May 17, 1991)

184 Jean M. Twenge and W. Keith Campbell, *The Narcissism Epidemic: Living in the Age of Entitlement* (New York: Free Press, A Division of Simon & Schuster, Inc. 2009)

185 Joseph Valadez and Remi Clignet, "On the Ambiguities of a Sociological Analysis of the Culture of Narcissism" *Sociological Quarterly*, vol. 28, 4 (Dec. 1987): 455–472

186 The Associated Press, "Isolationism soars among Americans" (March 12, 2009): http://www.msnbc.msn.com/id/34255911/ns/world_news/

187 "Poll: 44% Americans see China as top economic power," *People's Daily* (December 04, 2009), http://english.peopledaily.com.cn/90001/90776/90883/6831907.html

188 "A rising tide of hunger," *Los Angeles Times* (November 26, 2009): http://articles.latimes.com/2009/nov/26/opinion/la-ed-hunger26-2009nov26

189 Jason DeParle, "Hunger in U.S. at a 14-Year High," *The New York Times* (November 16, 2009), http://www.nytimes.com/2009/11/17/us/17hunger.html

190 Andrew Martin, "As Recession Deepens, So Does Milk Surplus, *The New York Times* (January 1, 2009), http://www.nytimes.com/2009/01/02/business/02dairy.html

191 Rav Avraham Kook (The Raaiah), *Igrot* (*Letters*), Vol. 2, 226

192 Ashlag, "Introduction to the Book, Panim Meirot uMasbirot" (Illuminating and Enlightening Face) in *Kabbalah for the Student*, 438

193 Johnson and Johnson, "Educational Psychology Success Story," 372

194 (ibid.)

195 (ibid.)

196 Johnson and Johnson, "Educational Psychology Success Story," 373

197 Johnson and Johnson, "Educational Psychology Success Story," 374

198 Jean M. Twenge, *Generation Me: Why Today's Young Americans Are More Confident, Assertive, Entitled--and More Miserable Than Ever Before* (USA: Free Press, March 6, 2007)

199 U.S. Department of Education, "Media Guide—Helping Your Child Through Early Adolescence," http://www2.ed.gov/parents/academic/help/adolescence/index.html

200 University of Michigan Health System, "Television and Children," http://www.med.umich.edu/yourchild/topics/tv.htm

201 (ibid.)

202 Barbara M. Newman and Philip R. Newman, *Development Through Life: A Psychosocial Approach* (Belmont, CA: Wadsworth Cengage Learning ,2008), 250

203 "Major Depression in Children and Adolescents," http://www.mentalhealthcanada.com/ConditionsandDisordersDetail.asp?lang=e&category=68

204 Ashlag, "The Peace," in *Kabbalah for the Student*, 273

ABOUT BNEI BARUCH

B nei Baruch is an international group of Kabbalists who share the wisdom of Kabbalah with the entire world. The study materials (in over 30 languages) are authentic Kabbalah texts that were passed down from generation to generation.

HISTORY AND ORIGIN

In 1991, following the passing of his teacher, Rav Baruch Shalom HaLevi Ashlag (The Rabash), Michael Laitman, Professor of Ontology and the Theory of Knowledge, PhD in Philosophy and Kabbalah, and MSc in Medical Bio-Cybernetics, established a Kabbalah study group called "Bnei Baruch." He called it Bnei Baruch (Sons of Baruch) to commemorate his mentor, whose side he never left in the final twelve years of his life, from 1979 to 1991. Dr. Laitman had been Ashlag's prime student and personal assistant, and is recognized as the successor to Rabash's teaching method.

The Rabash was the firstborn son and successor of Rav Yehuda Leib HaLevi Ashlag, the greatest Kabbalist of the 20th century. Rav Ashlag authored the most authoritative and comprehensive commentary on *The Book of Zohar*, titled *The Sulam* (Ladder) *Commentary*. He was the first to reveal the complete method for spiritual ascent, and thus was known as Baal HaSulam (Owner of the Ladder).

Bnei Baruch bases its entire study method on the path paved by these two great spiritual leaders.

THE STUDY METHOD

The unique study method developed by Baal HaSulam and his son, the Rabash, is taught and applied on a daily basis by Bnei Baruch. This method relies on authentic Kabbalah sources such as *The Book of Zohar*, by Rabbi Shimon Bar-Yochai, *The Tree of Life*, by the Ari, and *The Study of the Ten Sefirot*, by Baal HaSulam.

While the study relies on authentic Kabbalah sources, it is carried out in simple language and uses a scientific, contemporary approach. The unique combination of an academic study method and personal experiences broadens the students' perspective and awards them a new perception of the reality they live in. Those on the spiritual path are thus given the necessary tools to study themselves and their surrounding reality.

Bnei Baruch is a diverse movement of tens of thousands of students worldwide. Students can choose their own paths and intensity of their studies according to their unique conditions and abilities.

THE MESSAGE

The essence of the message disseminated by Bnei Baruch is universal: unity of the people, unity of nations and love of man.

For millennia, Kabbalists have been teaching that love of man should be the foundation of all human relations. This love prevailed in the days of Abraham, Moses, and the group of Kabbalists that they established. If we make room for these seasoned, yet contemporary values, we will discover that we possess the power to put differences aside and unite.

The wisdom of Kabbalah, hidden for millennia, has been waiting for the time when we would be sufficiently developed and ready to implement its message. Now, it is emerging as a solution that can unite diverse factions everywhere, enabling us, as individuals and as a society, to meet today's challenges.

ACTIVITIES

Bnei Baruch was established on the premise that "only by expansion of the wisdom of Kabbalah to the public can we be awarded complete redemption" (Baal HaSulam). Therefore, Bnei Baruch offers a variety of ways for people to explore and discover the purpose of their lives, providing careful guidance for beginners and advanced students alike.

Internet

Bnei Baruch's international website, www.kab.info, presents the authentic wisdom of Kabbalah using essays, books, and original texts. It is by far the most expansive source of authentic Kabbalah material on the Internet, containing a unique, extensive library for readers to thoroughly explore the wisdom of Kabbalah. Additionally, the media archive, www.kabbalahmedia.info, contains thousands of media items, downloadable books, and a vast reservoir of texts, video and audio files in many languages.

Bnei Baruch's online Kabbalah Education Center offers free Kabbalah courses for beginners, initiating students into this profound body of knowledge in the comfort of their own homes.

Dr. Laitman's daily lessons are also aired live on www.kab.tv, along with complementary texts and diagrams.

All these services are provided free of charge.

Television

In Israel, Bnei Baruch established its own channel, no. 66 on both cable and satellite, which broadcasts 24/7 Kabbalah TV. The channel is also aired on the Internet at www.kab.tv. All broadcasts on the channel are free of charge. Programs are adapted for all levels, from complete beginners to the most advanced.

Conferences

Twice a year, students gather for a weekend of study and socializing at conferences in various locations in the U.S., as well as an annual convention in Israel. These gatherings provide a great setting for meeting like-minded people, for bonding, and for expanding one's understanding of the wisdom.

Kabbalah Books

Bnei Baruch publishes authentic books, written by Baal HaSulam, his son, the Rabash, as well as books by Dr. Michael Laitman. The books of Rav Ashlag and Rabash are essential for complete understanding of the teachings of authentic Kabbalah, explained in Laitman's lessons.

Dr. Laitman writes his books in a clear, contemporary style based on the key concepts of Baal HaSulam. These books are a vital link between today's readers and the original texts. All the books are available for sale, as well as for free download.

Paper

Kabbalah Today is a free paper produced and disseminated by Bnei Baruch in many languages, including English, Hebrew,

Spanish, and Russian. It is apolitical, non-commercial, and written in a clear, contemporary style. The purpose of *Kabbalah Today* is to expose the vast knowledge hidden in the wisdom of Kabbalah at no cost and in a clear, engaging style for readers everywhere.

Kabbalah Lessons

As Kabbalists have been doing for centuries, Laitman gives a daily lesson. The lessons are given in Hebrew and are simultaneously interpreted into seven languages—English, Russian, Spanish, French, German, Italian, and Turkish—by skilled and experienced interpreters. As with everything else, the live broadcast is free of charge.

FUNDING

Bnei Baruch is a non-profit organization for teaching and sharing the wisdom of Kabbalah. To maintain its independence and purity of intentions, Bnei Baruch is not supported, funded, or otherwise tied to any government or political organization.

Since the bulk of its activity is provided free of charge, the prime sources of funding for the group's activities are donations and tithing—contributed by students on a voluntary basis—and Dr. Laitman's books, which are sold at cost.

HOW TO CONTACT
BNEI BARUCH

1057 Steeles Avenue West, Suite 532
Toronto, ON, M2R 3X1
Canada

Bnei Baruch USA,
2009 85th street, #51,
Brooklyn, New York, 11214
USA

E-mail: info@kabbalah.info
Web site: www.kabbalah.info

Toll free in USA and Canada:
1-866-LAITMAN
Fax: 1-905 886 9697

FURTHER READING

To help you determine which book you would like to read next, we have divided the books into six categories—Beginners, Intermediate, Advanced, Good for All, Textbooks, and For Children. The first three categories are divided by the level of prior knowledge readers are required to have in order to easily relate to the book. The Beginners Category requires no prior knowledge. The Intermediate Category requires reading one or two beginners' books first; and the Advanced level requires one or two books of each of the previous categories. The fourth category, Good for All, includes books you can always enjoy, whether you are a complete novice or well versed in Kabbalah.

The fifth category—Textbooks—includes translations of authentic source materials from earlier Kabbalists, such as the Ari, Rav Yehuda Ashlag (Baal HaSulam) and his son and successor, Rav Baruch Ashlag (the Rabash). As its name implies, the sixth category—For Children—includes books that are suitable for children ages 3 and above. Those are not Kabbalah books per se, but are rather inspired by the teaching and convey the Kabbalistic message of love and unity.

Additional material that has not yet been published can be found at www.kabbalah.info. All materials on this site, including e-versions of published books, can be downloaded free of charge directly from the store at www.kabbalahbooks.info.

BEGINNERS

A Guide to the Hidden Wisdom of Kabbalah

A Guide to the Hidden Wisdom of Kabbalah is a light and reader-friendly guide to beginners in Kabbalah, covering everything from the history of Kabbalah to how this wisdom can help resolve the world crisis.

The book is set up in three parts: Part 1 covers the history, facts, and fallacies about Kabbalah, and introduces its key concepts. Part 2 tells you all about the spiritual worlds and other neat stuff like the meaning of letters and the power of music. Part 3 covers the implementation of Kabbalah at a time of world crisis.

Kabbalah Revealed

This is the most clearly written, reader-friendly guide to making sense of the surrounding world. Each of its six chapters focuses on a different aspect of the wisdom of Kabbalah, illuminating its teachings and explaining them using various examples from our day-to-day lives.

The first three chapters in Kabbalah Revealed explain why the world is in a state of crisis, how our growing desires promote progress as well as alienation, and why the biggest deterrent to achieving positive change is rooted in our own spirits. Chapters Four through Six offer a prescription for positive change. In these chapters, we learn how we can use our spirits to build a personally peaceful life in harmony with all of Creation.

Wondrous Wisdom

This book offers an initial course on Kabbalah. Like all the books presented here, *Wondrous Wisdom* is based solely on authentic teachings passed down from Kabbalist teacher to student over thousands of years. At the heart of the book is a sequence of lessons revealing the nature of Kabbalah's wisdom and explaining how to attain it. For every person questioning "Who am I really?" and "Why am I on this planet?" this book is a must.

Awakening to Kabbalah

A distinctive, personal, and awe-filled introduction to an ancient wisdom tradition. In this book, Rav Laitman offers a deeper understanding of the fundamental teachings of Kabbalah, and how you can use its wisdom to clarify your relationship with others and the world around you.

Using language both scientific and poetic, he probes the most profound questions of spirituality and existence. This provocative, unique guide will inspire and invigorate you to see beyond the world as it is and the limitations of your everyday life, become closer to the Creator, and reach new depths of the soul.

Kabbalah, Science, and the Meaning of Life

Science explains the mechanisms that sustain life; Kabbalah explains why life exists. In *Kabbalah, Science, and the Meaning of Life*, Rav Laitman combines science and spirituality in a captivating dialogue that reveals life's meaning.

For thousands of years Kabbalists have been writing that the world is a single entity divided into separate beings. Today the cutting-edge science of quantum physics states a very similar idea: that at the most fundamental level of matter, we are all literally one.

Science proves that reality is affected by the observer who examines it; and so does Kabbalah. But Kabbalah makes an even bolder statement: even the Creator, the Maker of reality, is within the observer. In other words, God is inside of us; He doesn't exist anywhere else. When we pass away, so does He.

These earthshaking concepts and more are eloquently introduced so that even readers new to Kabbalah or science will easily understand them. Therefore, if you're just a little curious about why you are here, what life means, and what you can do to enjoy it more, this book is for you.

From Chaos to Harmony

Many researchers and scientists agree that the ego is the reason behind the perilous state our world is in today. Laitman's groundbreaking book not only demonstrates that egoism has been the basis for all suffering throughout human history, but also shows how we can turn our plight to pleasure.

The book contains a clear analysis of the human soul and its problems, and provides a "roadmap" of what we need to do to once again be happy. *From Chaos to Harmony* explains how we can rise to a new level of existence on personal, social, national, and international levels.

Kabbalah for Beginners

Kabbalah for Beginners is a book for all those seeking answers to life's essential questions. We all want to know why we are here, why there is pain, and how we can make life more enjoyable. The four parts of this book provide us with reliable answers to these questions, as well as clear explanations of the gist of Kabbalah and its practical implementations.

Part One discusses the discovery of the wisdom of Kabbalah, and how it was developed, and finally concealed until our time. Part Two introduces the gist of the wisdom of Kabbalah, using ten easy drawings to help us understand the structure of the spiritual worlds, and how they relate to our world. Part Three reveals Kabbalistic concepts that are largely unknown to the public, and Part Four elaborates on practical means you and I can take, to make our lives better and more enjoyable for us and for our children.

INTERMEDIATE

The Kabbalah Experience

The depth of the wisdom revealed in the questions and answers within this book will inspire readers to reflect and contemplate. This is not a book to race through, but rather one that should be read thoughtfully and carefully. With this approach, readers will begin to experience a growing sense of enlightenment while simply absorbing the answers to the questions every Kabbalah student asks along the way.

The Kabbalah Experience is a guide from the past to the future, revealing situations that all students of Kabbalah will experience at some point along their journeys. For those who cherish every moment in life, this book offers unparalleled insights into the timeless wisdom of Kabbalah.

The Path of Kabbalah

This unique book combines beginners' material with more advanced concepts and teachings. If you have read a book or two of Laitman's, you will find this book very easy to relate to.

While touching upon basic concepts such as perception of reality and Freedom of Choice, *The Path of Kabbalah* goes deeper and expands beyond the scope of beginners' books. The structure of the worlds, for example, is explained in greater detail here than in the "pure" beginners' books. Also described is the spiritual root of mundane matters such as the Hebrew calendar and the holidays.

ADVANCED

The Science of Kabbalah

Kabbalist and scientist Rav Michael Laitman, PhD, designed this book to introduce readers to the special language and terminology of the authentic wisdom of Kabbalah. Here, Rav Laitman reveals authentic Kabbalah in a manner both rational and mature. Readers are gradually led to understand the logical design of the Universe and the life that exists in it.

The Science of Kabbalah, a revolutionary work unmatched in its clarity, depth, and appeal to the intellect, will enable readers to approach the more technical works of Baal HaSulam (Rabbi Yehuda Ashlag), such as *The Study of the Ten Sefirot* and *The Book of Zohar*. Readers of this book will enjoy the satisfying answers to the riddles of life that only authentic Kabbalah provides. Travel through the pages and prepare for an astonishing journey into the Upper Worlds.

Introduction to the Book of Zohar

This volume, along with *The Science of Kabbalah*, is a required preparation for those who wish to understand the hidden message of *The Book of Zohar*. Among the many helpful topics dealt with in this text is an introduction to the "language of roots and branches," without which the stories in *The Zohar* are mere fable and legend. *Introduction to the Book of Zohar* will provide readers with the necessary tools to understand authentic Kabbalah as it was originally meant to be—as a means to attain the Upper Worlds.

The Book of Zohar: annotations to the Ashlag commentary

The Book of Zohar (*The Book of Radiance*) is an age-old source of wisdom and the basis for all Kabbalistic literature. Since its appearance nearly 2,000 years ago, it has been the primary, and often only, source used by Kabbalists.

For centuries, Kabbalah was hidden from the public, which was deemed not yet ready to receive it. However, our generation has been designated by Kabbalists as the first generation that *is* ready to grasp the concepts in *The Zohar*. Now we can put these principles into practice in our lives.

Written in a unique and metaphorical language, *The Book of Zohar* enriches our understanding of reality and widens our worldview. Although the text deals with one subject only—how to relate to the Creator—it approaches it from different angles. This allows each of us to find the particular phrase or word that will carry us into the depths of this profound and timeless wisdom.

GOOD FOR ALL

The Point in the Heart

The Point in the Heart; a Source of Delight for My Soul is a unique collection of excerpts from a man whose wisdom has earned him devoted students in North America and the world over. Michael Laitman is a scientist, a Kabbalist, and a great thinker who presents ancient wisdom in a compelling style.

This book does not profess to teach Kabbalah, but rather gently introduces ideas from the teaching. *The Point in the Heart* is a window to a new perception. As the author himself testifies to the wisdom of Kabbalah, "It is a science of emotion, a science of pleasure. You are welcome to open and to taste."

Attaining the Worlds Beyond

From the introduction to *Attaining the Worlds Beyond*: "...Not feeling well on the Jewish New Year's Eve of September 1991, my teacher called me to his bedside and handed me his notebook, saying, 'Take it and learn from it.' The following morning, he perished in my arms, leaving me and many of his other disciples without guidance in this world.

"He used to say, 'I want to teach you to turn to the Creator, rather than to me, because He is the only strength, the only Source of all that exists, the only one who can really help you, and He awaits your prayers for help. When you seek help in your search for freedom from the bondage of this world, help in elevating yourself above this world, help in finding the self, and help in determining your purpose in life, you must turn to the Creator, who sends you all those aspirations in order to compel you to turn to Him.'"

Attaining the Worlds Beyond holds within it the content of that notebook, as well as other inspiring texts. This book reaches out to all those seekers who want to find a logical, reliable way to understand the world's phenomena. This fascinating introduction to the wisdom of Kabbalah will enlighten the mind, invigorate the heart, and move readers to the depths of their souls.

Bail Yourself Out

In *Bail Yourself Out: how you can emerge strong from the world crisis*, Laitman introduces several extraordinary concepts that weave into a complete solution: 1) The crisis is essentially not financial, but *psychological*: People have stopped trusting each other, and where there is no trust there is no trade, but only war, isolation, and pain. 2) This mistrust is a result of a *natural process* that's been evolving for millennia and is culminating today. 3) To resolve the crisis, we must first *understand* the process that created the alienation. 4) The first, and most important, step to understanding the crisis is to *inform* people about this natural process through books, such as *Bail Yourself Out*, TV, cinema, and any other means of communication. 5) With this information, we will "*revamp*" our relationships and build them on trust, collaboration, and most importantly, care. This mending process will guarantee that we and our families will prosper in a world of plenty.

Basic Concepts in Kabbalah

This is a book to help readers cultivate an *approach to the concepts* of Kabbalah, to spiritual objects, and to spiritual terms. By reading and re-reading in this book, one develops internal observations, senses, and approaches that did not previously exist within. These newly acquired observations are like sensors that "feel" the space around us that is hidden from our ordinary senses.

Hence, *Basic Concepts in Kabbalah* is intended to foster the contemplation of spiritual terms. Once we are integrated with these terms, we can begin to see, with our inner vision, the unveiling of the spiritual structure that surrounds us, almost as if a mist has been lifted.

This book is not aimed at the study of facts. Instead, it is a book for those who wish to awaken the deepest and subtlest sensations they can possess.

Children of Tomorrow:
Guidelines for Raising Happy Children in the 21st Century

Children of Tomorrow is a new beginning for you and your children. Imagine being able to hit the reboot button and get it right this time. No hassle, no stress, and best of all—no guessing.

The big revelation is that raising kids is all about games and play, relating to them as small grownups, and making all major decisions together. You will be surprised to discover how teaching kids about positive things like friendship and caring for others automatically spills into other areas of our lives through the day.

Open any page and you will find thought-provoking quotes about every aspect of children's lives: parent-children relations, friendships and conflicts, and a clear picture of how schools should be designed and function. This book offers a fresh perspective on how to raise our children, with the goal being the happiness of all children everywhere.

The Wise Heart:
Tales and allegories by three contemporary sages

"Our inner work is to tune our hearts and our senses to perceive the spiritual world," says Michael Laitman in the poem Spiritual Wave. *The Wise Heart* is a lovingly crafted anthology comprised of tales and allegories by Kabbalist Dr. Michael Laitman, his mentor, Rav Baruch Ashlag (Rabash), and Rabash's father and mentor, Rav Yehuda Ashlag, author of the acclaimed *Sulam* (Ladder) commentary on *The Book of Zohar*.

Kabbalah students and enthusiasts in Kabbalah often wonder what the spiritual world actually feels like to a Kabbalist. The allegories in this delicate compilation provide a glimpse into those feelings.

The poems herein are excerpts from letters and lessons given by these three spiritual giants to their students through the years. They offer surprising and often amusing depictions of human nature, with a loving and tender touch that is truly unique to Kabbalists. Indeed, *The Wise Heart* is a gift of wisdom and delight for any wisdom seeking heart.

TEXTBOOKS

Shamati

Rav Michael Laitman's words on the book: Among all the texts and notes that were used by my teacher, Rav Baruch Shalom Halevi Ashlag (the Rabash), there was one special notebook he always carried. This notebook contained the transcripts of his conversations with his father, Rav Yehuda Leib Halevi Ashlag (Baal HaSulam), author of the *Sulam* (Ladder) commentary on *The Book of Zohar*, *The Study of the Ten Sefirot* (a commentary on the texts of the Kabbalist, Ari), and of many other works on Kabbalah.

Not feeling well on the Jewish New Year's Eve of September 1991, the Rabash summoned me to his bedside and handed me a notebook, whose cover contained only one word, *Shamati* (I Heard). As he handed the notebook, he said, "Take it and learn from it." The following morning, my teacher perished in my arms, leaving me and many of his other disciples without guidance in this world.

Committed to Rabash's legacy to disseminate the wisdom of Kabbalah, I published the notebook just as it was written, thus retaining the text's transforming powers. Among all the books of Kabbalah, *Shamati* is a unique and compelling creation.

Kabbalah for the Student

Kabbalah for the Student offers authentic texts by Rav Yehuda Ashlag, author of the *Sulam* (Ladder) commentary on *The Book of Zohar*, his son and successor, Rav Baruch Ashlag, as well as other great Kabbalists. It also offers illustrations that accurately depict the evolution of the Upper Worlds as Kabbalists experience them. The book also contains several explanatory essays that help us understand the texts within.

In *Kabbalah for the Student*, Rav Michael Laitman, PhD, Rav Baruch Ashlag's personal assistant and prime student, compiled all the texts a Kabbalah student would need in order to attain the spiritual worlds. In his daily lessons, Rav Laitman bases his teaching on these inspiring texts, thus helping novices and veterans alike to better understand the spiritual path we undertake on our fascinating journey to the Higher Realms.

Rabash—the Social Writings

Rav Baruch Shalom HaLevi Ashlag (Rabash) played a remarkable role in the history of Kabbalah. He provided us with the necessary final link connecting the wisdom of Kabbalah to our human experience. His father and teacher was the great Kabbalist, Rav Yehuda Leib HaLevi Ashlag, known as Baal HaSulam for his *Sulam* (Ladder) commentary on *The Book of Zohar*. Yet, if not for the essays of Rabash, his father's efforts to disclose the wisdom of Kabbalah to all would have been in vain. Without those essays, few would be able to achieve the spiritual attainment that Baal HaSulam so desperately wanted us to obtain.

The writings in this book aren't just for reading. They are more like an experiential user's guide. It is very important to work with them in order to see what they truly contain. The reader should try to put them into practice by living out the emotions Rabash so masterfully describes. He always advised his students to summarize the articles, to work with the texts, and those who attempt it discover that it always yields new insights. Thus, readers are advised to work with the texts, summarize them, translate them, and implement them in the group. Those who do so will discover the power in the writings of Rabash.

Gems of Wisdom:
words of the great Kabbalists from all generations

Through the millennia, Kabbalists have bequeathed us with numerous writings. In their compositions, they have laid out a structured method that can lead, step by step, unto a world of eternity and wholeness.

Gems of wisdom is a collection of selected excerpts from the writings of the greatest Kabbalists from all generations, with particular emphasis on the writings of Rav Yehuda Leib HaLevi Ashlag (Baal HaSulam), author of the *Sulam* [Ladder] commentary of *The Book of Zohar*.

The sections have been arranged by topics, to provide the broadest view possible on each topic. This book is a useful guide to any person desiring spiritual advancement.

FOR CHILDREN

Together Forever

On the surface, *Together Forever* is a children's story. But like all good children's stories, it transcends boundaries of age, culture, and upbringing.

In *Together Forever*, the author tells us that if we are patient and endure the trials we encounter along our life's path, we will become stronger, braver, and wiser. Instead of growing weaker, we will learn to create our own magic and our own wonders as only a magician canIn this warm, tender tale, Michael Laitman shares with children and parents alike some of the gems and charms of the spiritual world. The wisdom of Kabbalah is filled with spellbinding stories. *Together Forever* is yet another gift from this ageless source of wisdom, whose lessons make our lives richer, easier, and far more fulfilling.

Miracles Can Happen

"Miracles Can Happen," Princes Peony," and "Mary and the Paints" are only three of ten beautiful stories for children ages 3-10. Written especially for children, these short tales convey a single message of love, unity, and care for all beings. The unique illustrations were carefully crafted to contribute to the overall message of the book, and a child who's heard or read any story in this collection is guaranteed to go to sleep smiling.

The Baobab that Opened Its Heart: and Other Nature Tales for Children

The Baobab that Opened Its Heart is a collection of stories for children, but not just for them. The stories in this collection were written with the love of nature, of people, and specifically with children in mind. They all share the desire to tell nature's tale of unity, connectedness, and love.

Kabbalah teaches that love is nature's guiding force, the reason for creation. The stories in this book convey it in the unique way that Kabbalah engenders in its students. The variety of authors and diversity of styles allows each reader to find the story that they like most.